Sustainable Energy

Sustainable Energy

Edited by
Ivy Stanford

⊟ Larsen & Keller
www.larsen-keller.com

Sustainable Energy
Edited by Ivy Stanford
ISBN: 978-1-63549-140-1 (Hardback)

© 2017 Larsen & Keller

Larsen & Keller

Published by Larsen and Keller Education,
5 Penn Plaza,
19th Floor,
New York, NY 10001, USA

Cataloging-in-Publication Data

Sustainable energy / edited by Ivy Stanford.
       p. cm.
Includes bibliographical references and index.
ISBN 978-1-63549-140-1
    1. Renewable energy sources. 2. Clean energy. 3. Renewable energy sources--Technological innovations.
I. Stanford, Ivy.
TJ808 .S87 2017
621.042--dc23

The publisher's policy is to use permanent paper from mills that operate a sustainable forestry policy. Furthermore, the publisher ensures that the text paper and cover boards used have met acceptable environmental accreditation standards.

Printed and bound in the United States of America.

For more information regarding Larsen and Keller Education and its products, please visit the publisher's website www.larsen-keller.com

Table of Contents

Preface

Sustainable energy refers to the science of producing electricity from sustainable and renewable resources to minimize pollution. The resources used for this practice are biomass power, solar power, tidal power, wind power, geothermal power, hydropower, nuclear power, etc. The aim of this book is to familiarize readers with this vast and rapidly growing subject of sustainable energy. It is designed to provide the students in-depth knowledge about this subject. The topics included in this extensive textbook are of utmost importance and are bound to provide thorough insights to the reader about this field. Some of the diverse topics covered in it address the varied branches that fall under this category. This text is an essential guide for students in the field of renewable energy systems engineering, environmental monitoring and analysis and power engineering.

A detailed account of the significant topics covered in this book is provided below:

Chapter 1- Green energy differs from other alternative energy sources as it focuses to provide cleaner options for fuel consumption as well as being a long-term and constant source of energy. The chapter on green energy offers an insightful focus, keeping in mind the complex subject matter.

Chapter 2- Renewable energy has recently become very popular in providing energy to the world and providing jobs. It ensures the slowing down of climate change as well as provides energy security. Discussed in this chapter are the major sources of power such as solar power, hydropower, wind power etc.

Chapter 3- Constant research and technical development has led to the adoption of green energy compatible technology in machinery like cars and buses as well as in the design of buildings and recreational centres. The chapter strategically encompasses and incorporates the major components and key concepts of green energy, providing a complete understanding.

Chapter 4- With more and more machinery relying on green energy technologies, greater effort has been made to provide efficient and adaptable devices. Some of them that are discussed in this chapter are solar cell, wind turbine and biomass briquettes. This chapter discusses the methods of green energy technologies in a critical manner providing key analysis to the subject matter.

It gives me an immense pleasure to thank our entire team for their efforts. Finally in the end, I would like to thank my family and colleagues who have been a great source of inspiration and support.

Editor

Introduction to Green Energy

Green energy differs from other alternative energy sources as it focuses to provide cleaner options for fuel consumption as well as being a long-term and constant source of energy. The chapter on green energy offers an insightful focus, keeping in mind the complex subject matter.

Sustainable energy is energy that is consumed at insignificant rates compared to its supply and with manageable collateral effects, especially environmental effects. Another common definition of sustainable energy is an energy system that serves the needs of the present without compromising the ability of future generations to meet their needs. The organizing principle for sustainability is sustainable development, which includes the four interconnected domains: ecology, economics, politics and culture. Sustainability science is the study of sustainable development and environmental science.

Technologies that promote sustainable energy include renewable energy sources, such as hydroelectricity, solar energy, wind energy, wave power, geothermal energy, bioenergy, tidal power and also technologies designed to improve energy efficiency. Costs have fallen dramatically in recent years, and continue to fall. Most of these technologies are either economically competitive or close to being so. Increasingly, effective government policies support investor confidence and these markets are expanding. Considerable progress is being made in the energy transition from fossil fuels to ecologically sustainable systems, to the point where many studies support 100% renewable energy.

Definitions

Energy efficiency and renewable energy are said to be the *twin pillars* of sustainable energy. Some ways in which *sustainable energy* has been defined are:

- "Effectively, the provision of energy such that it meets the needs of the present without compromising the ability of future generations to meet their own needs. ...Sustainable Energy has two key components: renewable energy and energy efficiency." – *Renewable Energy and Efficiency Partnership* (British)

- "Dynamic harmony between equitable availability of energy-intensive goods and services to all people and the preservation of the earth for future generations." And, "The solution will lie in finding sustainable energy sources and more efficient means of converting and utilizing energy." – *Sustainable Energy* by J. W. Tester, *et al.*, from MIT Press.

"Any energy generation, efficiency and conservation source where: Resources are available to enable massive scaling to become a significant portion of energy generation, long term, preferably 100 years.." – *Invest,* a green technology non-profit organization.

"Energy which is replenishable within a human lifetime and causes no long-term damage to the environment." – *Jamaica Sustainable Development Network*

This sets *sustainable energy* apart from other renewable energy terminology such as *alternative energy* by focusing on the ability of an energy source to continue providing energy. Sustainable energy can produce some pollution of the environment, as long as it is not sufficient to prohibit heavy use of the source for an indefinite amount of time. Sustainable energy is also distinct from low-carbon energy, which is sustainable only in the sense that it does not add to the CO_2 in the atmosphere.

Green Energy is energy that can be extracted, generated, and/or consumed without any significant negative impact to the environment. The planet has a natural capability to recover which means pollution that does not go beyond that capability can still be termed green.

Green power is a subset of renewable energy and represents those renewable energy resources and technologies that provide the highest environmental benefit. The U.S. Environmental Protection Agency defines green power as electricity produced from solar, wind, geothermal, biogas, biomass and low-impact small hydroelectric sources. Customers often buy green power for avoided environmental impacts and its greenhouse gas reduction benefits.

Renewable Energy Technologies

Renewable energy technologies are essential contributors to sustainable energy as they generally contribute to world energy security, reducing dependence on fossil fuel resources, and providing opportunities for mitigating greenhouse gases. The International Energy Agency states that:

Conceptually, one can define three generations of renewables technologies, reaching back more than 100 years .

First-generation technologies emerged from the industrial revolution at the end of the 19th century and include hydropower, biomass combustion and geothermal power and heat. Some of these technologies are still in widespread use.

Second-generation technologies include solar heating and cooling, wind power, modern forms of bioenergy and solar photovoltaics. These are now entering markets as a result of research, development and demonstration (RD&D) investments since the 1980s. The initial investment was prompted by energy security concerns linked to the oil crises (1973 and 1979) of the 1970s but the continuing appeal of these renewables is due, at least in part, to environmental benefits. Many of the technologies reflect significant advancements in materials.

Third-generation technologies are still under development and include advanced biomass gasification, biorefinery technologies, concentrating solar thermal power, hot dry rock geothermal energy and ocean energy. Advances in nanotechnology may also play a major role.

—*International Energy Agency, RENEWABLES IN GLOBAL ENERGY SUPPLY, An IEA Fact Sheet*

First- and second-generation technologies have entered the markets, and third-generation technologies heavily depend on long term research and development commitments, where the public sector has a role to play.

Regarding energy used by vehicles, a comprehensive 2008 cost-benefit analysis review was conducted of sustainable energy sources and usage combinations in the context of global warming and other dominating issues; it ranked wind power generation combined with battery electric vehicles (BEV) and hydrogen fuel cell vehicles (HFCVs) as the most efficient. Wind was followed by concentrated solar power (CSP), geothermal power, tidal power, photovoltaic, wave power, hydropower coal capture and storage (CCS), nuclear energy and biofuel energy sources. It states: "In sum, use of wind, CSP, geothermal, tidal, PV, wave, and hydro to provide electricity for BEVs and HFCVs and, by extension, electricity for the residential, industrial, and commercial sectors, will result in the most benefit among the options considered. The combination of these technologies should be advanced as a solution to global warming, air pollution, and energy security. Coal-CCS and nuclear offer less benefit thus represent an opportunity cost loss, and the biofuel options provide no certain benefit and the greatest negative impacts."

First-generation Technologies

One of many power plants at The Geysers, a geothermal power field in northern California, with a total output of over 750 MW.

First-generation technologies are most competitive in locations with abundant resources. Their future use depends on the exploration of the available resource potential, particularly in developing countries, and on overcoming challenges related to the environment and social acceptance.

—International Energy Agency, RENEWABLES IN GLOBAL ENERGY SUPPLY, An IEA Fact Sheet

Among sources of renewable energy, hydroelectric plants have the advantages of being long-lived— many existing plants have operated for more than 100 years. Also, hydroelectric plants are clean and have few emissions. Criticisms directed at large-scale hydroelectric plants include: dislocation of people living where the reservoirs are planned, and release of significant amounts of carbon dioxide during construction and flooding of the reservoir.

Hydroelectric dams are one of the most widely deployed sources of sustainable energy.

However, it has been found that high emissions are associated only with shallow reservoirs in warm (tropical) locales, and recent innovations in hydropower turbine technology are enabling efficient development of low-impact run-of-the-river hydroelectricity projects. Generally speaking, hydroelectric plants produce much lower life-cycle emissions than other types of generation. Hydroelectric power, which underwent extensive development during growth of electrification in the 19th and 20th centuries, is experiencing resurgence of development in the 21st century. The areas of greatest hydroelectric growth are the booming economies of Asia. China is the development leader; however, other Asian nations are installing hydropower at a rapid pace. This growth is driven by much increased energy costs—especially for imported energy—and widespread desires for more domestically produced, clean, renewable, and economical generation.

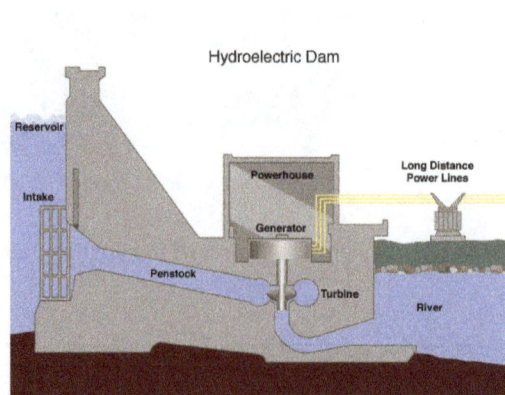

Hydroelectric dam in cross section

Geothermal power plants can operate 24 hours per day, providing base-load capacity, and the world potential capacity for geothermal power generation is estimated at 85 GW over the next 30 years. However, geothermal power is accessible only in limited areas of the world, including the United States, Central America, East Africa, Iceland, Indonesia, and the Philippines. The costs of geothermal energy have dropped substantially from the systems built in the 1970s. Geothermal heat generation can be competitive in many countries producing geothermal power, or in other regions where the resource is of a lower temperature. Enhanced geothermal system (EGS) technology does not require natural convective hydrothermal resources, so it can be used in areas that were previously unsuitable for geothermal power, if the resource is very large. EGS is currently under research at the U.S. Department of Energy.

Biomass briquettes are increasingly being used in the developing world as an alternative to charcoal. The technique involves the conversion of almost any plant matter into compressed briquettes that typically have about 70% the calorific value of charcoal. There are relatively few examples of large-scale briquette production. One exception is in North Kivu, in eastern Democratic Republic of Congo, where forest clearance for charcoal production is considered to be the biggest threat to mountain gorilla habitat. The staff of Virunga National Park have successfully trained and equipped over 3500 people to produce biomass briquettes, thereby replacing charcoal produced illegally inside the national park, and creating significant employment for people living in extreme poverty in conflict-affected areas.

Second-generation Technologies

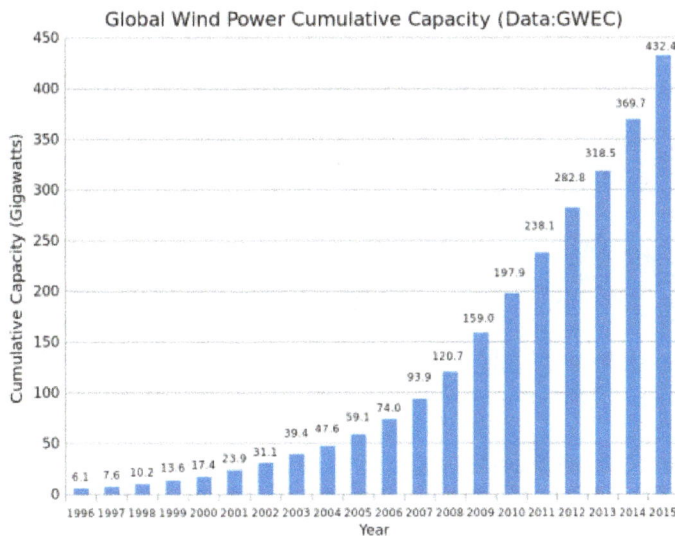

Wind power: worldwide installed capacity

Markets for second-generation technologies are strong and growing, but only in a few countries. The challenge is to broaden the market base for continued growth worldwide. Strategic deployment in one country not only reduces technology costs for users there, but also for those in other countries, contributing to overall cost reductions and performance improvement.

—*International Energy Agency, RENEWABLES IN GLOBAL ENERGY SUPPLY, An IEA Fact Sheet*

Solar heating systems are a well known second-generation technology and generally consist of solar thermal collectors, a fluid system to move the heat from the collector to its point of usage, and a reservoir or tank for heat storage and subsequent use. The systems may be used to heat domestic hot water, swimming pool water, or for space heating. The heat can also be used for industrial applications or as an energy input for other uses such as cooling equipment. In many climates, a solar heating system can provide a very high percentage (50 to 75%) of domestic hot water energy. Energy received from the sun by the earth is that of electromagnetic radiation. Light ranges of visible, infrared, ultraviolet, x-rays, and radio waves received by the earth through solar energy. The highest power of radiation comes from visible light. Solar power is complicated due to changes in seasons and from day to night. Cloud cover can also add to complications of solar energy, and not all radiation from the sun reaches earth because it is absorbed and dispersed due to clouds and gases within the earth's atmospheres.

11 MW solar power plant near Serpa, Portugal 38°1′51″N 7°37′22″W

In the 1980s and early 1990s, most photovoltaic modules provided remote-area power supply, but from around 1995, industry efforts have focused increasingly on developing building integrated photovoltaics and power plants for grid connected applications. Currently the largest photovoltaic power plant in North America is the Nellis Solar Power Plant (15 MW). There is a proposal to build a Solar power station in Victoria, Australia, which would be the world's largest PV power station, at 154 MW. Other large photovoltaic power stations include the Girassol solar power plant (62 MW), and the Waldpolenz Solar Park (40 MW).

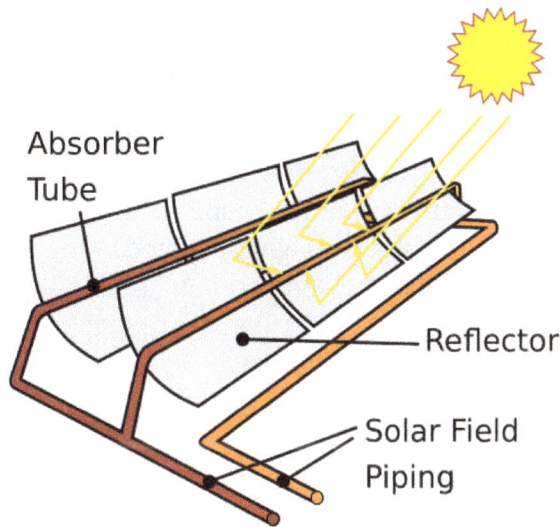

Sketch of a Parabolic Trough Collector

Some of the second-generation renewables, such as wind power, have high potential and have already realised relatively low production costs. At the end of 2008, worldwide wind farm capacity was 120,791 megawatts (MW), representing an increase of 28.8 percent during the year, and wind power produced some 1.3% of global electricity consumption. Wind power accounts for approximately 20% of electricity use in Denmark, 9% in Spain, and 7% in Germany. However, it may be difficult to site wind turbines in some areas for aesthetic or environmental reasons, and it may be difficult to integrate wind power into electricity grids in some cases.

Solar thermal power stations have been successfully operating in California commercially since the late 1980s, including the largest solar power plant of any kind, the 350 MW Solar Energy Generating Systems. Nevada Solar One is another 64MW plant which has recently opened. Other parabolic trough power plants being proposed are two 50MW plants in Spain, and a 100MW plant in Israel.

Information on pump, California

Brazil has one of the largest renewable energy programs in the world, involving production of ethanol fuel from sugar cane, and ethanol now provides 18 percent of the country's automotive fuel. As a result of this, together with the exploitation of domestic deep water oil sources, Brazil, which years ago had to import a large share of the petroleum needed for domestic consumption, recently reached complete self-sufficiency in oil.

Most cars on the road today in the U.S. can run on blends of up to 10% ethanol, and motor vehicle manufacturers already produce vehicles designed to run on much higher ethanol blends. Ford, DaimlerChrysler, and GM are among the automobile companies that sell "flexible-fuel" cars, trucks, and minivans that can use gasoline and ethanol blends ranging from pure gasoline up to 85% ethanol (E85). By mid-2006, there were approximately six million E85-compatible vehicles on U.S. roads.

Third-generation Technologies

MIT's Solar House#1 built in 1939 used seasonal thermal energy storage (STES) for year-round heating.

Third-generation technologies are not yet widely demonstrated or commercialised. They are on the horizon and may have potential comparable to other renewable energy technologies, but still depend on attracting sufficient attention and RD&D funding. These newest technologies include advanced biomass gasification, biorefinery technologies, solar thermal power stations, hot dry rock geothermal energy and ocean energy.

—*International Energy Agency, RENEWABLES IN GLOBAL ENERGY SUPPLY, An IEA Fact Sheet*

Bio-fuels may be defined as "renewable," yet may not be "sustainable," due to soil degradation. As of 2012, 40% of American corn production goes toward ethanol. Ethanol takes up a large percentage of "Clean Energy Use" when in fact, it is still debatable whether ethanol should be considered as a "Clean Energy."

According to the International Energy Agency, new bioenergy (biofuel) technologies being developed today, notably cellulosic ethanol biorefineries, could allow biofuels to play a much bigger role in the future than previously thought. Cellulosic ethanol can be made from plant matter composed primarily of inedible cellulose fibers that form the stems and branches of most plants. Crop residues (such as corn stalks, wheat straw and rice straw), wood waste and municipal solid waste are potential sources of cellulosic biomass. Dedicated energy crops, such as switchgrass, are also promising cellulose sources that can be sustainably produced in many regions of the United States.

The world's first commercial tidal stream generator – SeaGen – in Strangford Lough. The strong wake shows the power in the tidal current.

In terms of ocean energy, another third-generation technology, Portugal has the world's first commercial wave farm, the Aguçadora Wave Park, under construction in 2007. The farm will initially use three Pelamis P-750 machines generating 2.25 MW. and costs are put at 8.5 million euro. Subject to successful operation, a further 70 million euro is likely to be invested before 2009 on a further 28 machines to generate 525 MW. Funding for a wave farm in Scotland was announced in February, 2007 by the Scottish Executive, at a cost of over 4 million pounds, as part of a £13 million funding packages for ocean power in Scotland. The farm will be the world's largest with a capacity of 3 MW generated by four Pelamis machines.

In 2007, the world's first turbine to create commercial amounts of energy using tidal power was installed in the narrows of Strangford Lough in Ireland. The 1.2 MW underwater tidal electricity generator takes advantage of the fast tidal flow in the lough which can be up to 4m/s. Although the generator is powerful enough to power up to a thousand homes, the turbine has a minimal environmental impact, as it is almost entirely submerged, and the rotors turn slowly enough that they pose no danger to wildlife.

Solar power panels that use nanotechnology, which can create circuits out of individual silicon molecules, may cost half as much as traditional photovoltaic cells, according to executives and investors involved in developing the products. Nanosolar has secured more than $100 million from investors to build a factory for nanotechnology thin-film solar panels. The company's plant has a planned production capacity of 430 megawatts peak power of solar cells per year. Commercial production started and first panels have been shipped to customers in late 2007.

Large national and regional research projects on artificial photosynthesis are designing nanotechnology-based systems that use solar energy to split water into hydrogen fuel. and a proposal has been made for a Global Artificial Photosynthesis project In 2011, researchers at the Massachusetts

Institute of Technology (MIT) developed what they are calling an "Artificial Leaf", which is capable of splitting water into hydrogen and oxygen directly from solar power when dropped into a glass of water. One side of the "Artificial Leaf" produces bubbles of hydrogen, while the other side produces bubbles of oxygen.

Most current solar power plants are made from an array of similar units where each unit is continuously adjusted, e.g., with some step motors, so that the light converter stays in focus of the sun light. The cost of focusing light on converters such as high-power solar panels, Stirling engine, etc. can be dramatically decreased with a simple and efficient rope mechanics. In this technique many units are connected with a network of ropes so that pulling two or three ropes is sufficient to keep all light converters simultaneously in focus as the direction of the sun changes.

Japan and China have national programs aimed at commercial scale Space-Based Solar Power (SBSP). The China Academy of Space Technology (CAST) won the 2015 International SunSat Design Competition with this video of their Multi-Rotary Joint design. Proponents of SBSP claim that Space-Based Solar Power would be clean, constant, and global, and could scale to meet all planetary energy demand. A recent multi-agency industry proposal (echoing the 2008 Pentagon recommendation) won the SECDEF/SECSTATE/USAID Director D3 (Diplomacy, Development, Defense) Innovation Challenge.

Enabling Technologies for Renewable Energy

Heat pumps and Thermal energy storage are classes of technologies that can enable the utilization of renewable energy sources that would otherwise be inaccessible due to a temperature that is too low for utilization or a time lag between when the energy is available and when it is needed. While enhancing the temperature of available renewable thermal energy, heat pumps have the additional property of leveraging electrical power (or in some cases mechanical or thermal power) by using it to extract additional energy from a low quality source (such as seawater, lake water, the ground, the air, or waste heat from a process).

Thermal storage technologies allow heat or cold to be stored for periods of time ranging from hours or overnight to interseasonal, and can involve storage of sensible energy (i.e. by changing the temperature of a medium) or latent energy (i.e. through phase changes of a medium, such between water and slush or ice). Short-term thermal storages can be used for peak-shaving in district heating or electrical distribution systems. Kinds of renewable or alternative energy sources that can be enabled include natural energy (e.g. collected via solar-thermal collectors, or dry cooling towers used to collect winter's cold), waste energy (e.g. from HVAC equipment, industrial processes or power plants), or surplus energy (e.g. as seasonally from hydropower projects or intermittently from wind farms). The Drake Landing Solar Community (Alberta, Canada) is illustrative. borehole thermal energy storage allows the community to get 97% of its year-round heat from solar collectors on the garage roofs, which most of the heat collected in summer. Types of storages for sensible energy include insulated tanks, borehole clusters in substrates ranging from gravel to bedrock, deep aquifers, or shallow lined pits that are insulated on top. Some types of storage are capable of storing heat or cold between opposing seasons (particularly if very large), and some storage applications require inclusion of a heat pump. Latent heat is typically stored in ice tanks or what are called phase-change materials (PCMs).

Energy Efficiency

Moving towards energy sustainability will require changes not only in the way energy is supplied, but in the way it is used, and reducing the amount of energy required to deliver various goods or services is essential. Opportunities for improvement on the demand side of the energy equation are as rich and diverse as those on the supply side, and often offer significant economic benefits.

Renewable energy and energy efficiency are sometimes said to be the "twin pillars" of sustainable energy policy. Both resources must be developed in order to stabilize and reduce carbon dioxide emissions. Efficiency slows down energy demand growth so that rising clean energy supplies can make deep cuts in fossil fuel use. If energy use grows too fast, renewable energy development will chase a receding target. A recent historical analysis has demonstrated that the rate of energy efficiency improvements has generally been outpaced by the rate of growth in energy demand, which is due to continuing economic and population growth. As a result, despite energy efficiency gains, total energy use and related carbon emissions have continued to increase. Thus, given the thermodynamic and practical limits of energy efficiency improvements, slowing the growth in energy demand is essential. However, unless clean energy supplies come online rapidly, slowing demand growth will only begin to reduce total emissions; reducing the carbon content of energy sources is also needed. Any serious vision of a sustainable energy economy thus requires commitments to both renewables and efficiency.

Renewable energy (and energy efficiency) are no longer niche sectors that are promoted only by governments and environmentalists. The increased levels of investment and the fact that much of the capital is coming from more conventional financial actors suggest that sustainable energy options are now becoming mainstream. An example of this would be The Alliance to Save Energy's Project with Stahl Consolidated Manufacturing, (Huntsville, Alabama, USA) (StahlCon 7), a patented generator shaft designed to reduce emissions within existing power generating systems, granted publishing rights to the Alliance in 2007.

Climate change concerns coupled with high oil prices and increasing government support are driving increasing rates of investment in the sustainable energy industries, according to a trend analysis from the United Nations Environment Programme. According to UNEP, global investment in sustainable energy in 2007 was higher than previous levels, with $148 billion of new money raised in 2007, an increase of 60% over 2006. Total financial transactions in sustainable energy, including acquisition activity, was $204 billion.

Investment flows in 2007 broadened and diversified, making the overall picture one of greater breadth and depth of sustainable energy use. The mainstream capital markets are "now fully receptive to sustainable energy companies, supported by a surge in funds destined for clean energy investment".

Smart-grid Technology

Smart grid refers to a class of technology people are using to bring utility electricity delivery systems into the 21st century, using computer-based remote control and automation. These systems are made possible by two-way communication technology and computer processing that has been used for decades in other industries. They are beginning to be used on electricity networks, from the power plants and wind farms all the way to the consumers of electricity in homes and businesses. They offer many benefits to utilities and consumers—mostly seen in big improvements in energy efficiency on the electricity grid and in the energy users' homes and offices.

Green Energy and Green Power

A solar trough array is an example of green energy

Green energy includes natural energetic processes that can be harnessed with little pollution. Green power is electricity generated from renewable energy sources.

Public seat with integrated solar panel in Singapore. Anyone can sit and plug in their mobile phone for free charging

Anaerobic digestion, geothermal power, wind power, small-scale hydropower, solar energy, biomass power, tidal power, wave power, and some forms of nuclear power (ones which are able to "burn" nuclear waste through a process known as nuclear transmutation, such as an Integral Fast Reactor, and therefore belong in the "Green Energy" category). Some definitions may also include power derived from the incineration of waste.

Some people, including Greenpeace founder and first member Patrick Moore, George Monbiot, Bill Gates and James Lovelock have specifically classified nuclear power as green energy. Others, including Greenpeace's Phil Radford disagree, claiming that the problems associated with radioactive waste and the risk of nuclear accidents (such as the Chernobyl disaster) pose an unacceptable risk to the environment and to humanity. However, newer nuclear reactor designs are capable of utilizing what is now deemed "nuclear waste" until it is no longer (or dramatically less) dangerous, and have design features that greatly minimize the possibility of a nuclear accident. These designs have yet to be proven.

Some have argued that although green energy is a commendable effort in solving the world's increasing energy consumption, it must be accompanied by a cultural change that encourages the decrease of the world's appetite for energy.

In several countries with common carrier arrangements, electricity retailing arrangements make it possible for consumers to purchase green electricity (renewable electricity) from either their utility or a green power provider.

When energy is purchased from the electricity network, the power reaching the consumer will not necessarily be generated from green energy sources. The local utility company, electric company, or state power pool buys their electricity from electricity producers who may be generating from fossil fuel, nuclear or renewable energy sources. In many countries green energy currently provides a very small amount of electricity, generally contributing less than 2 to 5% to the overall pool. In some U.S. states, local governments have formed regional power purchasing pools using Community Choice Aggregation and Solar Bonds to achieve a 51% renewable mix or higher, such as in the City of San Francisco.

By participating in a green energy program a consumer may be having an effect on the energy sources used and ultimately might be helping to promote and expand the use of green energy. They are also making a statement to policy makers that they are willing to pay a price premium to support renewable energy. Green energy consumers either obligate the utility companies to increase the amount of green energy that they purchase from the pool (so decreasing the amount of non-green energy they purchase), or directly fund the green energy through a green power provider. If insufficient green energy sources are available, the utility must develop new ones or contract with a third party energy supplier to provide green energy, causing more to be built. However, there is no way the consumer can check whether or not the electricity bought is "green" or otherwise.

In some countries such as the Netherlands, electricity companies guarantee to buy an equal amount of 'green power' as is being used by their green power customers. The Dutch government exempts green power from pollution taxes, which means green power is hardly any more expensive than other power.

In the United States, one of the main problems with purchasing green energy through the electrical grid is the current centralized infrastructure that supplies the consumer's electricity. This infrastructure has led to increasingly frequent brown outs and black outs, high CO_2 emissions, higher energy costs, and power quality issues. An additional $450 billion will be invested to expand this fledgling system over the next 20 years to meet increasing demand. In addition, this centralized system is now being further overtaxed with the incorporation of renewable energies such as wind, solar, and geothermal energies. Renewable resources, due to the amount of space they require, are often located in remote areas where there is a lower energy demand. The current infrastructure would make transporting this energy to high demand areas, such as urban centers, highly inefficient and in some cases impossible. In addition, despite the amount of renewable energy produced or the economic viability of such technologies only about 20 percent will be able to be incorporated into the grid. To have a more sustainable energy profile, the United States must move towards implementing changes to the electrical grid that will accommodate a mixed-fuel economy.

However, several initiatives are being proposed to mitigate these distribution problems. First and foremost, the most effective way to reduce USA's CO_2 emissions and slow global warming is through conservation efforts. Opponents of the current US electrical grid have also advocated for decentralizing the grid. This system would increase efficiency by reducing the amount of energy lost in transmission. It would also be economically viable as it would reduce the amount of power lines that will need to be constructed in the future to keep up with demand. Merging heat and power in this system would create added benefits and help to increase its efficiency by up to 80-90%. This is a significant increase from the current fossil fuel plants which only have an efficiency of 34%.

A more recent concept for improving our electrical grid is to beam microwaves from Earth-orbiting satellites or the moon to directly when and where there is demand. The power would be generated from solar energy captured on the lunar surface In this system, the receivers would be "broad, translucent tent-like structures that would receive microwaves and convert them to electricity". NASA said in 2000 that the technology was worth pursuing but it is still too soon to say if the technology will be cost-effective.

The World Wide Fund for Nature and several green electricity labelling organizations created the (now defunct) Eugene Green Energy Standard under which the national green electricity certification schemes could be accredited to ensure that the purchase of green energy leads to the provision of additional new green energy resources.

Local Green Energy Systems

A small Quietrevolution QR5 Gorlov type vertical axis wind turbine in Bristol, England. Measuring 3 m in diameter and 5 m high, it has a nameplate rating of 6.5 kW to the grid.

Those not satisfied with the third-party grid approach to green energy via the power grid can install their own locally based renewable energy system. Renewable energy electrical systems from solar to wind to even local hydro-power in some cases, are some of the many types of renewable energy systems available locally. Additionally, for those interested in heating and cooling their dwelling via renewable energy, geothermal heat pump systems that tap the constant temperature of the earth, which is around 7 to 15 degrees Celsius a few feet underground and increases dramatically at greater depths, are an option over conventional natural gas and petroleum-fueled heat approaches. Also, in geographic locations where the Earth's Crust is especially thin, or near volcanoes (as is the case in Iceland) there exists the potential to generate even more electricity than would be possible at other sites, thanks to a more significant temperature gradient at these locales.

The advantage of this approach in the United States is that many states offer incentives to offset the cost of installation of a renewable energy system. In California, Massachusetts and several other U.S. states, a new approach to community energy supply called Community Choice Aggregation has provided communities with the means to solicit a competitive electricity supplier and use municipal revenue bonds to finance development of local green energy resources. Individuals are usually assured that the electricity they are using is actually produced from a green energy source that they control. Once the system is paid for, the owner of a renewable energy system will be producing their own renewable electricity for essentially no cost and can sell the excess to the local utility at a profit.

Using Green Energy

A 01 KiloWatt Micro Windmill for Domestic Usage

Renewable energy, after its generation, needs to be stored in a medium for use with autonomous devices as well as vehicles. Also, to provide household electricity in remote areas (that is areas which are not connected to the mains electricity grid), energy storage is required for use with renewable energy. Energy generation and consumption systems used in the latter case are usually stand-alone power systems.

Some examples are:

- energy carriers as hydrogen, liquid nitrogen, compressed air, oxyhydrogen, batteries, to power vehicles.

- flywheel energy storage, pumped-storage hydroelectricity is more usable in stationary applications (e.g. to power homes and offices). In household power systems, conversion of energy can also be done to reduce smell. For example, organic matter such as cow dung and spoilable organic matter can be converted to biochar. To eliminate emissions, carbon capture and storage is then used.

Usually however, renewable energy is derived from the mains electricity grid. This means that energy storage is mostly not used, as the mains electricity grid is organised to produce the exact amount of energy being consumed at that particular moment. Energy production on the mains electricity grid is always set up as a combination of (large-scale) renewable energy plants, as well as other power plants as fossil-fuel power plants and nuclear power. This combination however, which is essential for this type of energy supply (as e.g. wind turbines, solar power plants etc.) can only produce when the wind blows and the sun shines. This is also one of the main drawbacks of the system as fossil fuel powerplants are polluting and are a main cause of global warming (nuclear power being an exception). Although fossil fuel power plants too can be made emissionless (through carbon capture and storage), as well as renewable (if the plants are converted to e.g. biomass) the best solution is still to phase out the latter power plants over time. Nuclear power plants too can be more or less eliminated from their problem of nuclear waste through the use of nuclear reprocessing and newer plants as fast breeder and nuclear fusion plants.

Renewable energy power plants do provide a steady flow of energy. For example, hydropower plants, ocean thermal plants, osmotic power plants all provide power at a regulated pace, and are thus available power sources at any given moment (even at night, windstill moments etc.). At present however, the number of steady-flow renewable energy plants alone is still too small to meet energy demands at the times of the day when the irregular producing renewable energy plants cannot produce power.

Besides the greening of fossil fuel and nuclear power plants, another option is the distribution and immediate use of power from solely renewable sources. In this set-up energy storage is again not necessary. For example, TREC has proposed to distribute solar power from the Sahara to Europe. Europe can distribute wind and ocean power to the Sahara and other countries. In this way, power is produced at any given time as at any point of the planet as the sun or the wind is up or ocean waves and currents are stirring. This option however is probably not possible in the short-term, as fossil fuel and nuclear power are still the main sources of energy on the mains electricity net and replacing them will not be possible overnight.

Several large-scale energy storage suggestions for the grid have been done. Worldwide there is over 100 GW of Pumped-storage hydroelectricity. This improves efficiency and decreases energy losses but a conversion to an energy storing mains electricity grid is a very costly solution. Some costs could potentially be reduced by making use of energy storage equipment the consumer buys and not the state. An example is batteries in electric cars that would double as an energy buffer for the electricity grid. However besides the cost, setting-up such a system would still be a very complicated and difficult procedure. Also, energy storage apparatus' as car batteries are also built with materials that pose a threat to the environment (e.g. Lithium). The combined production of batteries for such a large part of the population would still have environmental concerns. Besides car batteries however, other Grid energy storage projects make use of less polluting energy carriers (e.g. compressed air tanks and flywheel energy storage).

Carbon-neutral and Negative Fuels

A carbon-neutral fuel is a synthetic fuel – such as methane, gasoline, diesel fuel or jet fuel – produced from renewable or nuclear energy used to hydrogenate waste carbon dioxide recycled from power plant flue-gas emissions, recovered from automotive exhaust gas, or derived from carbonic acid in seawater. Such fuels are carbon-neutral because they do not result in a net increase in atmospheric greenhouse gases. To the extent that carbon-neutral fuels displace fossil fuels, or if they are produced from waste carbon or seawater carbonic acid, and their combustion is subject to carbon capture at the flue or exhaust pipe, they result in negative carbon dioxide emission and net carbon dioxide removal from the atmosphere, and thus constitute a form of greenhouse gas remediation. Such fuels are produced by the electrolysis of water to make hydrogen used in turn in the Sabatier reaction to produce methane which may then be stored to be burned later in power plants as synthetic natural gas, transported by pipeline, truck, or tanker ship, or be used in gas to liquids processes such as the Fischer–Tropsch process to make traditional transportation or heating fuels.

Green Energy and Labeling by Region

European Union

Directive 2004/8/EC of the European Parliament and of the Council of 11 February 2004 on the promotion of cogeneration based on a useful heat demand in the internal energy market includes the article 5 (*Guarantee of origin of electricity* from high-efficiency cogeneration).

European environmental NGOs have launched an ecolabel for green power. The ecolabel is called EKOenergy. It sets criteria for sustainability, additionality, consumer information and tracking. Only part of electricity produced by renewables fulfills the EKOenergy criteria.

A Green Energy Supply Certification Scheme was launched in the United Kingdom in February 2010. This implements guidelines from the Energy Regulator, Ofgem, and sets requirements on transparency, the matching of sales by renewable energy supplies, and additionality.

United States

The United States Department of Energy (DOE), the Environmental Protection Agency (EPA), and the Center for Resource Solutions (CRS) recognizes the voluntary purchase of electricity from renewable energy sources (also called renewable electricity or green electricity) as green power.

The most popular way to purchase renewable energy as revealed by NREL data is through purchasing Renewable Energy Certificates (RECs). According to a Natural Marketing Institute (NMI) survey 55 percent of American consumers want companies to increase their use of renewable energy.

DOE selected six companies for its 2007 Green Power Supplier Awards, including Constellation NewEnergy; 3Degrees; Sterling Planet; SunEdison; Pacific Power and Rocky Mountain Power; and Silicon Valley Power. The combined green power provided by those six winners equals more than 5 billion kilowatt-hours per year, which is enough to power nearly 465,000 average U.S. households. In 2014, Arcadia Power made RECS available to homes and businesses in all 50 states, allowing consumers to use "100% green power" as defined by the EPA's Green Power Partnership.

The U.S. Environmental Protection Agency (USEPA) Green Power Partnership is a voluntary program that supports the organizational procurement of renewable electricity by offering expert advice, technical support, tools and resources. This can help organizations lower the transaction costs of buying renewable power, reduce carbon footprint, and communicate its leadership to key stakeholders.

Throughout the country, more than half of all U.S. electricity customers now have an option to purchase some type of green power product from a retail electricity provider. Roughly one-quarter of the nation's utilities offer green power programs to customers, and voluntary retail sales of renewable energy in the United States totaled more than 12 billion kilowatt-hours in 2006, a 40% increase over the previous year.

Sustainable Energy Research

There are numerous organizations within the academic, federal, and commercial sectors conducting large scale advanced research in the field of sustainable energy. This research spans several areas of focus across the sustainable energy spectrum. Most of the research is targeted at improving efficiency and increasing overall energy yields. Multiple federally supported research organizations have focused on sustainable energy in recent years. Two of the most prominent of these labs are Sandia National Laboratories and the National Renewable Energy Laboratory (NREL), both of which are funded by the United States Department of Energy and supported by various corporate partners. Sandia has a total budget of $2.4 billion while NREL has a budget of $375 million.

Scientific production towards sustainable energy systems is rising exponentially, growing from about 500 English journal papers only about renewable energy in 1992 to almost 9,000 papers in 2011.

Solar

The primary obstacle that is preventing the large scale implementation of solar powered energy generation is the inefficiency of current solar technology. Currently, photovoltaic (PV) panels only have the ability to convert around 16% of the sunlight that hits them into electricity. At this rate, many experts believe that solar energy is not efficient enough to be economically sustainable given the cost to produce the panels themselves. Both Sandia National Laboratories and the National Renewable Energy Laboratory (NREL), have heavily funded solar research programs. The NREL solar program has a budget of around $75 million and develops research projects in the areas of photovoltaic (PV) technology, solar thermal energy, and solar radiation. The budget for Sandia's solar division is unknown, however it accounts for a significant percentage of the laboratory's $2.4 billion budget. Several academic programs have focused on solar research in recent years. The Solar Energy Research Center (SERC) at University of North Carolina (UNC) has the sole purpose of developing cost effective solar technology. In 2008, researchers at Massachusetts Institute of Technology (MIT) developed a method to store solar energy by using it to produce hydrogen fuel from water. Such research is targeted at addressing the obstacle that solar development faces of storing energy for use during nighttime hours when the sun is not shining. In February 2012, North Carolina-based Semprius Inc., a solar development company backed by German corporation Siemens, announced that they had developed the world's most efficient solar panel. The company claims that the prototype converts 33.9% of the sunlight that hits it to electricity, more than double the previous high-end conversion rate. Major projects on artificial photosynthesis or solar fuels are also under way in many developed nations.

Space-Based Solar Power

Space-Based Solar Power Satellites seek to overcome the problems of storage and provide civilization-scale power that is clean, constant, and global. Japan and China have active national programs aimed at commercial scale Space-Based Solar Power (SBSP), and both nation's hope to orbit demonstrations in the 2030s. The China Academy of Space Technology (CAST) won the 2015 International SunSat Design Competition with this video of their Multi-Rotary Joint design. Proponents of SBSP claim that Space-Based Solar Power would be clean, constant, and global, and could scale to meet all planetary energy demand. A recent multi-agency industry proposal (echoing the 2008 Pentagon recommendation) won the SECDEF/SECSTATE/USAID Director D3 (Diplomacy, Development, Defense) Innovation Challenge with the following pitch and vision video. Northrop Grumman is funding CALTECH with $17.5 million for an ultra lightweight design. Keith Henson recently posted a video of a "bootstrapping" approach.

Wind

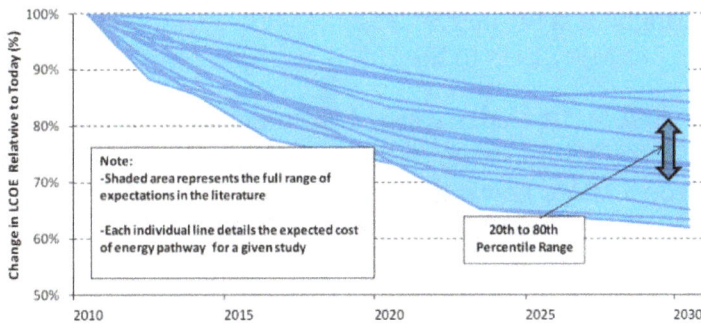

The National Renewable Energy Laboratory projects that the levelized cost of wind power in the U.S. will decline about 25% from 2012 to 2030.

Bangui Wind Farm in the Philippines.

Wind energy research dates back several decades to the 1970s when NASA developed an analytical model to predict wind turbine power generation during high winds. Today, both Sandia National Laboratories and National Renewable Energy Laboratory have programs dedicated to wind research. Sandia's laboratory focuses on the advancement of materials, aerodynamics, and sensors. The NREL wind projects are centered on improving wind plant power production, reducing their capital costs, and making wind energy more cost effective overall. The Field Laboratory for Optimized Wind Energy (FLOWE) at Caltech was established to research renewable approaches to wind energy farming technology practices that have the potential to reduce the cost, size, and environmental impact of wind energy production. The president of Sky WindPower Corporation thinks that wind turbines will be able to produce electricity at a cent/kWh at an average which in comparison to coal-generated electricity is a fractional of the cost.

A wind farm is a group of wind turbines in the same location used to produce electric power. A large wind farm may consist of several hundred individual wind turbines, and cover an extended area of hundreds of square miles, but the land between the turbines may be used for agricultural or other purposes. A wind farm may also be located offshore.

Many of the largest operational onshore wind farms are located in the USA and China. The Gansu Wind Farm in China has over 5,000 MW installed with a goal of 20,000 MW by 2020. China has several other "wind power bases" of similar size. The Alta Wind Energy Center in California is the largest onshore wind farm outside of China, with a capacity of 1020 MW of power. Europe leads in the use of wind power with almost 66 GW, about 66 percent of the total globally, with Denmark in the lead according to the countries installed per-capita capacity. As of February 2012, the Walney Wind Farm in United Kingdom is the largest offshore wind farm in the world at 367 MW, followed by Thanet Wind Farm (300 MW), also in the UK.

There are many large wind farms under construction and these include BARD Offshore 1 (400 MW), Clyde Wind Farm (350 MW), Greater Gabbard wind farm (500 MW), Lincs Wind Farm (270 MW), London Array (1000 MW), Lower Snake River Wind Project (343 MW), Macarthur Wind Farm (420 MW), Shepherds Flat Wind Farm (845 MW), and Sheringham Shoal (317 MW).

Wind power has expanded quickly, it's share of worldwide electricity usage at the end of 2014 was 3.1%.

Carbon-neutral and Negative Fuels

Carbon-neutral fuels are synthetic fuels (including methane, gasoline, diesel fuel, jet fuel or ammonia) produced by hydrogenating waste carbon dioxide recycled from power plant flue-gas emissions, recovered from automotive exhaust gas, or derived from carbonic acid in seawater. Commercial fuel synthesis companies suggest they can produce synthetic fuels for less than petroleum fuels when oil costs more than $55 per barrel. Renewable methanol (RM) is a fuel produced from hydrogen and carbon dioxide by catalytic hydrogenation where the hydrogen has been obtained from water electrolysis. It can be blended into transportation fuel or processed as a chemical feedstock.

The George Olah carbon dioxide recycling plant operated by Carbon Recycling International in Grindavík, Iceland has been producing 2 million liters of methanol transportation fuel per year from flue exhaust of the Svartsengi Power Station since 2011. It has the capacity to produce 5 million liters per year. A 250 kilowatt methane synthesis plant was constructed by the Center for Solar Energy and Hydrogen Research (ZSW) at Baden-Württemberg and the Fraunhofer Society in Germany and began operating in 2010. It is being upgraded to 10 megawatts, scheduled for completion in autumn, 2012. Audi has constructed a carbon-neutral liquefied natural gas (LNG) plant in Werlte, Germany. The plant is intended to produce transportation fuel to offset LNG used in their A3 Sportback g-tron automobiles, and can keep 2,800 metric tons of CO_2 out of the environment per year at its initial capacity. Other commercial developments are taking place in Columbia, South Carolina, Camarillo, California, and Darlington, England.

Such fuels are considered carbon-neutral because they do not result in a net increase in atmospheric greenhouse gases. To the extent that synthetic fuels displace fossil fuels, or if they are produced from waste carbon or seawater carbonic acid, and their combustion is subject to carbon capture at the flue or exhaust pipe, they result in negative carbon dioxide emission and net carbon dioxide removal from the atmosphere, and thus constitute a form of greenhouse gas remediation.

Such renewable fuels alleviate the costs and dependency issues of imported fossil fuels without requiring either electrification of the vehicle fleet or conversion to hydrogen or other fuels, enabling continued compatible and affordable vehicles. Carbon-neutral fuels offer relatively low cost energy storage, alleviating the problems of wind and solar intermittency, and they enable distribution of wind, water, and solar power through existing natural gas pipelines.

Nighttime wind power is considered the most economical form of electrical power with which to synthesize fuel, because the load curve for electricity peaks sharply during the warmest hours of the day, but wind tends to blow slightly more at night than during the day, so, the price of nighttime wind power is often much less expensive than any alternative. Germany has built a 250 kilowatt synthetic methane plant which they are scaling up to 10 megawatts.

Biomass

Sugarcane plantation to produce ethanol in Brazil

Biomass is biological material derived from living, or recently living organisms. It most often refers to plants or plant-derived materials which are specifically called lignocellulosic biomass. As an energy source, biomass can either be used directly via combustion to produce heat, or indirectly after converting it to various forms of biofuel. Conversion of biomass to biofuel can be achieved by different methods which are broadly classified into: *thermal*, *chemical*, and *biochemical* methods.

Wood remains the largest biomass energy source today; examples include forest residues – such as dead trees, branches and tree stumps –, yard clippings, wood chips and even municipal solid waste. In the second sense, biomass includes plant or animal matter that can be converted into fibers or other industrial chemicals, including biofuels. Industrial biomass can be grown from numerous types of plants, including miscanthus, switchgrass, hemp, corn, poplar, willow, sorghum, sugarcane, bamboo, and a variety of tree species, ranging from eucalyptus to oil palm (palm oil).

A CHP power station using wood to provide electricity to over 30.000 households in France

Biomass, biogas and biofuels are burned to produce heat/power and in doing so harm the environment. Pollutants such as sulphurous oxides (SO_x), nitrous oxides (NO_x), and particulate matter (PM) are produced from this combustion; the World Health Organisation estimates that 7 million premature deaths are caused each year by air pollution. Biomass combustion is a major contributor.

Ethanol Biofuels

As the primary source of biofuel in North America, many organizations are conducting research in the area of ethanol production. On the Federal level, the USDA conducts a large amount of research regarding ethanol production in the United States. Much of this research is targeted towards the effect of ethanol production on domestic food markets. The National Renewable Energy Laboratory has conducted various ethanol research projects, mainly in the area of cellulosic ethanol. Cellulosic ethanol has many benefits over traditional corn based-ethanol. It does not take away or directly conflict with the food supply because it is produced from wood, grasses, or non-edible parts of plants. Moreover, some studies have shown cellulosic ethanol to be more cost effective and economically sustainable than corn-based ethanol. Even if we used all the corn crop that we have in the United States and converted it into ethanol it would only produce enough fuel to serve 13 percent of the United States total gasoline consumption.Sandia National Laboratories conducts in-house cellulosic ethanol research and is also a member of the Joint BioEnergy Institute (JBEI), a research institute founded by the United States Department of Energy with the goal of developing cellulosic biofuels.

Other Biofuels

From 1978 to 1996, the National Renewable Energy Laboratory experimented with producing algae fuel in the "Aquatic Species Program." A self-published article by Michael Briggs, at the University of New Hampshire Biofuels Group, offers estimates for the realistic replacement of all motor vehicle fuel with biofuels by utilizing algae that have a natural oil content greater than 50%, which Briggs suggests can be grown on algae ponds at wastewater treatment plants. This oil-rich algae can then be extracted from the system and processed into biofuels, with the dried remainder further reprocessed to create ethanol. The production of algae to harvest oil for biofuels has not yet been undertaken on a commercial scale, but feasibility studies have been conducted to arrive at the above yield estimate. During the biofuel production process algae actually consumes the carbon dioxide in the air and turns it into oxygen through photosynthesis. In addition to its projected high yield, algaculture— unlike food crop-based biofuels — does not entail a decrease in food production, since it requires neither farmland nor fresh water. Many companies are pursuing algae bio-reactors for various purposes, including scaling up biofuels production to commercial levels.

Several groups in various sectors are conducting research on Jatropha curcas, a poisonous shrub-like tree that produces seeds considered by many to be a viable source of biofuels feedstock oil. Much of this research focuses on improving the overall per acre oil yield of Jatropha through advancements in genetics, soil science, and horticultural practices. SG Biofuels, a San Diego-based Jatropha developer, has used molecular breeding and biotechnology to produce elite hybrid seeds of Jatropha that show significant yield improvements over first generation varieties. The Center for Sustainable Energy Farming (CfSEF) is a Los Angeles-based non-profit research organization dedicated to Jatropha research in the areas of plant science, agronomy, and horticulture. Successful exploration of these disciplines is projected to increase Jatropha farm production yields by 200-300% in the next ten years.

Geothermal

Geothermal energy is produced by tapping into the thermal energy created and stored within the earth. It arises from the radioactive decay of an isotope of potassium and other elements found in the Earth's crust. Geothermal energy can be obtained by drilling into the ground, very similar to oil exploration, and then it is carried by a heat-transfer fluid (e.g. water, brine or steam). Geothermal systems that are mainly dominated by water have the potential to provide greater benefits to the system and will generate more power. Within these liquid-dominated systems, there are possible concerns of subsidence and contamination of ground-water resources. Therefore, protection of ground-water resources is necessary in these systems. This means that careful reservoir production and engineering is necessary in liquid-dominated geothermal reservoir systems. Geothermal energy is considered sustainable because that thermal energy is constantly replenished. However, the science of geothermal energy generation is still young and developing economic viability. Several entities, such as the National Renewable Energy Laboratory and Sandia National Laboratories are conducting research toward the goal of establishing a proven science around geothermal energy. The International Centre for Geothermal Research (IGC), a German geosciences research organization, is largely focused on geothermal energy development research.

Hydrogen

Over $1 billion of federal money has been spent on the research and development of hydrogen and a medium for energy storage in the United States. Both the National Renewable Energy Laboratory and Sandia National Laboratories have departments dedicated to hydrogen research. Hydrogen is useful for energy storage and for use in airplanes, but is not practical for automobile use, as it is not very efficient, compared to using a battery — for the same cost a person can travel three times as far using a battery.

Thorium

There are potentially two sources of nuclear power. Fission is used in all current nuclear power plants. Fusion is the reaction that exists in stars, including the sun, and remains impractical for use on Earth, as fusion reactors are not yet available. However nuclear power is controversial politically and scientifically due to concerns about radioactive waste disposal, safety, the risks of a severe accident, and technical and economical problems in dismantling of old power plants.

Thorium is a fissionable material used in thorium-based nuclear power. The thorium fuel cycle claims several potential advantages over a uranium fuel cycle, including greater abundance, superior physical and nuclear properties, better resistance to nuclear weapons proliferation and reduced plutonium and actinide production. Therefore, it is sometimes referred as sustainable.

Clean Energy Investments

2010 was a record year for green energy investments. According to a report from Bloomberg New Energy Finance, nearly US $243 billion was invested in wind farms, solar power, electric cars, and other alternative technologies worldwide, representing a 30 percent increase from 2009 and nearly five times the money invested in 2004. China had $51.1 billion investment in clean energy projects in 2010, by far the largest figure for any country.

Within emerging economies, Brazil comes second to China in terms of clean energy investments. Supported by strong energy policies, Brazil has one of the world's highest biomass and small-hydro power capacities and is poised for significant growth in wind energy investment. The cumulative investment potential in Brazil from 2010 to 2020 is projected as $67 billion.

India is another rising clean energy leader. While India ranked the 10th in private clean energy investments among G-20 members in 2009, over the next 10 years it is expected to rise to the third position, with annual clean energy investment under current policies forecast to grow by 369 percent between 2010 and 2020.

It is clear that the center of growth has started to shift to the developing economies and they may lead the world in the new wave of clean energy investments.

Around the world many sub-national governments - regions, states and provinces - have aggressively pursued sustainable energy investments. In the United States, California's leadership in renewable energy was recognised by The Climate Group when it awarded former Governor Arnold Schwarzenegger its inaugural award for international climate leadership in Copenhagen in 2009. In Australia, the state of South Australia - under the leadership of former Premier Mike

Rann - has led the way with wind power comprising 26% of its electricity generation by the end of 2011, edging out coal fired generation for the first time. South Australia also has had the highest take-up per capita of household solar panels in Australia following the Rann Government's introduction of solar feed-in laws and educative campaign involving the installation of solar photovoltaic installations on the roofs of prominent public buildings, including the parliament, museum, airport and Adelaide Showgrounds pavilion and schools. Rann, Australia's first climate change minister, passed legislation in 2006 setting targets for renewable energy and emissions cuts, the first legislation in Australia to do so.

Also, in the European Union there is a clear trend of promoting policies encouraging investments and financing for sustainable energy in terms of energy efficiency, innovation in energy exploitation and development of renewable resources, with increased consideration of environmental aspects and sustainability.

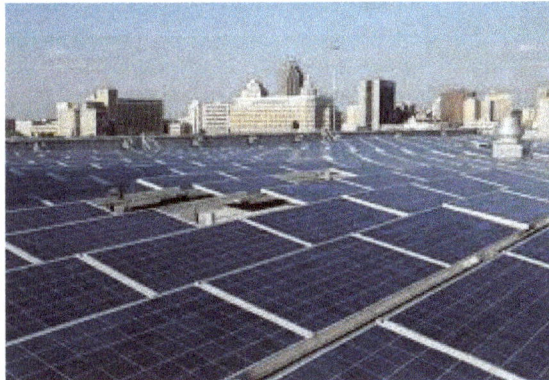

References

- Lynn R. Kahle, Eda Gurel-Atay, Eds (2014). Communicating Sustainability for the Green Economy. New York: M.E. Sharpe. ISBN 978-0-7656-3680-5.

- Huesemann, Michael H., and Joyce A. Huesemann (2011). Technofix: Why Technology Won't Save Us or the Environment, Chapter 5, "In Search of Solutions: Efficiency Improvements", New Society Publishers, ISBN 978-0-86571-704-6.

- "Th-ING: A Sustainable Energy Source | National Security Science Magazine | Los Alamos National Laboratory". lanl.gov. 2015. Retrieved 1 March 2015.

- Farah, Paolo Davide (2015). "Sustainable Energy Investments and National Security: Arbitration and Negotiation Issues". Journal of World Energy Law and Business. 8 (6). Retrieved 26 November 2015.

- Wong, Bill (28 June 2011), "Drake Landing Solar Community", IDEA/CDEA District Energy/CHP 2011 Conference, Toronto, pp. 1–30, retrieved 21 April 2013

- "Greenpeace International: The Founders (March 2007)". Web.archive.org. Archived from the original on 2007-02-03. Retrieved 2013-08-21.

- Monbiot, George (2009-02-20). "George Monbiot: A kneejerk rejection of nuclear power is not an option | Environment". London: theguardian.com. Retrieved 2013-08-21.

- "Climate Change as a Cultural and Behavioral Issue: Addressing Barriers and Implementing Solutions" (PDF). ScienceDirect. 2010. Retrieved 2013-08-28.

- Okulski, Travis (26 June 2012). "Audi's Carbon Neutral E-Gas Is Real And They're Actually Making It". Jalopnik (Gawker Media). Retrieved 29 July 2013.

- MacDowell, Niall; et al. (2010). "An overview of CO_2 capture technologies". Energy and Environmental Science. 3 (11): 1645–69. doi:10.1039/C004106H. Retrieved 7 September 2012.

- Goeppert, Alain; Czaun, Miklos; Prakash, G.K. Surya; Olah, George A. (2012). "Air as the renewable carbon source of the future: an overview of CO_2 capture from the atmosphere". Energy and Environmental Science. 5 (7): 7833–53. doi:10.1039/C2EE21586A. Retrieved 7 September 2012. (Review.)

- House, K.Z.; Baclig, A.C.; Ranjan, M.; van Nierop, E.A.; Wilcox, J.; Herzog, H.J. (2011). "Economic and energetic analysis of capturing CO_2 from ambient air" (PDF). Proceedings of the National Academy of Sciences. 108 (51): 20428–33. doi:10.1073/pnas.1012253108. Retrieved 7 September 2012. (Review.)

- Pennline, Henry W.; et al. (2010). "Separation of CO_2 from flue gas using electrochemical cells". Fuel. 89 (6): 1307–14. doi:10.1016/j.fuel.2009.11.036. Retrieved 7 September 2012.

Renewable Energy: An Overview

Renewable energy has recently become very popular in providing energy to the world and providing jobs. It ensures the slowing down of climate change as well as provides energy security. Discussed in this chapter are the major sources of power such as solar power, hydropower, wind power etc.

Renewable Energy

Wind, solar, and hydroelectricity are three emerging renewable sources of energy

Renewable energy is generally defined as energy that is collected from resources which are naturally replenished on a human timescale, such as sunlight, wind, rain, tides, waves, and geothermal heat. Renewable energy often provides energy in four important areas: electricity generation, air and water heating/cooling, transportation, and rural (off-grid) energy services.

Based on REN21's 2016 report, renewables contributed 19.2% to humans' global energy consumption and 23.7% to their generation of electricity in 2014 and 2015, respectively. This energy consumption is divided as 8.9% coming from traditional biomass, 4.2% as heat energy (modern biomass, geothermal and solar heat), 3.9% hydro electricity and 2.2% is electricity from wind, solar, geothermal, and biomass. Worldwide investments in renewable technologies amounted to more than US$286 billion in 2015, with countries like China and the United States heavily investing in wind, hydro, solar and biofuels. Globally, there are an estimated 7.7 million jobs associated with the renewable energy industries, with solar photovoltaics being the largest renewable employer.

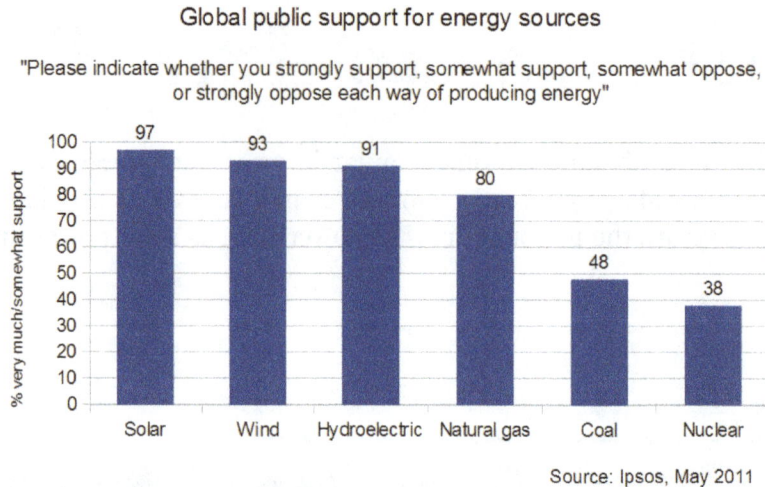

Global public support for energy sources

"Please indicate whether you strongly support, somewhat support, somewhat oppose, or strongly oppose each way of producing energy"

Energy Source	% very much/somewhat support
Solar	97
Wind	93
Hydroelectric	91
Natural gas	80
Coal	48
Nuclear	38

Source: Ipsos, May 2011

Global public support for different energy sources (2011) based on a poll by Ipsos Global @dvisor

Renewable energy resources exist over wide geographical areas, in contrast to other energy sources, which are concentrated in a limited number of countries. Rapid deployment of renewable energy and energy efficiency is resulting in significant energy security, climate change mitigation, and economic benefits. The results of a recent review of the literature concluded that as greenhouse gas (GHG) emitters begin to be held liable for damages resulting from GHG emissions resulting in climate change, a high value for liability mitigation would provide powerful incentives for deployment of renewable energy technologies. In international public opinion surveys there is strong support for promoting renewable sources such as solar power and wind power. At the national level, at least 30 nations around the world already have renewable energy contributing more than 20 percent of energy supply. National renewable energy markets are projected to continue to grow strongly in the coming decade and beyond. Some places and at least two countries, Iceland and Norway generate all their electricity using renewable energy already, and many other countries have the set a goal to reach 100% renewable energy in the future. For example, in Denmark the government decided to switch the total energy supply (electricity, mobility and heating/cooling) to 100% renewable energy by 2050.

While many renewable energy projects are large-scale, renewable technologies are also suited to rural and remote areas and developing countries, where energy is often crucial in human development. United Nations' Secretary-General Ban Ki-moon has said that renewable energy has the ability to lift the poorest nations to new levels of prosperity. As most of renewables provide electricity, renewable energy deployment is often applied in conjunction with further electrification, which has several benefits: For example, electricity can be converted to heat without losses and even reach higher temperatures than fossil fuels, can be converted into mechanical energy with high efficiency and is clean at the point of consumpion. In addition to that electrification with renewable energy is much more efficient and therefore leads to a significant reduction in primary energy requirements, because most renewables don't have a steam cycle with high losses (fossil power plants usually have losses of 40 to 65%).

Overview

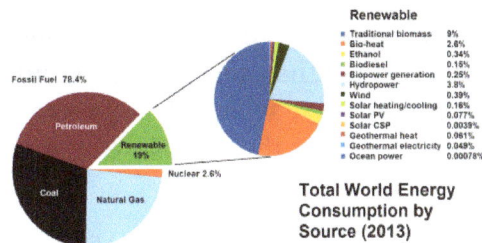

Renewable	
Traditional biomass	9%
Bio-heat	2.8%
Ethanol	0.34%
Biodiesel	0.16%
Biopower generation	0.25%
Hydropower	3.8%
Wind	0.39%
Solar heating/cooling	0.16%
Solar PV	0.077%
Solar CSP	0.0039%
Geothermal heat	0.081%
Geothermal electricity	0.049%
Ocean power	0.00078%

Total World Energy Consumption by Source (2013)

World energy consumption by source. Renewables accounted for 19% in 2012.

PlanetSolar, the world's largest solar-powered boat and the first ever solar electric vehicle to circumnavigate the globe (in 2012)

Renewable energy flows involve natural phenomena such as sunlight, wind, tides, plant growth, and geothermal heat, as the International Energy Agency explains:

Renewable energy is derived from natural processes that are replenished constantly. In its various forms, it derives directly from the sun, or from heat generated deep within the earth. Included in the definition is electricity and heat generated from solar, wind, ocean, hydropower, biomass, geothermal resources, and biofuels and hydrogen derived from renewable resources.

Renewable energy resources and significant opportunities for energy efficiency exist over wide geographical areas, in contrast to other energy sources, which are concentrated in a limited number of countries. Rapid deployment of renewable energy and energy efficiency, and technological diversification of energy sources, would result in significant energy security and economic benefits. It would also reduce environmental pollution such as air pollution caused by burning of fossil fuels and improve public health, reduce premature mortalities due to pollution and save associated health costs that amount to several hundred billion dollars annually only in the United States. Renewable energy sources, that derive their energy from the sun, either directly or indirectly, such as hydro and wind, are expected to be capable of supplying humanity energy for almost another 1 billion years, at which point the predicted increase in heat from the sun is expected to make the surface of the earth too hot for liquid water to exist.

Climate change and global warming concerns, coupled with high oil prices, peak oil, and increasing government support, are driving increasing renewable energy legislation, incentives and commercialization. New government spending, regulation and policies helped the industry weather the global financial crisis better than many other sectors. According to a 2011 projection by the International Energy Agency, solar power generators may produce most of the world's electricity within 50 years, reducing the emissions of greenhouse gases that harm the environment.

As of 2011, small solar PV systems provide electricity to a few million households, and micro-hydro configured into mini-grids serves many more. Over 44 million households use biogas made in household-scale digesters for lighting and/or cooking, and more than 166 million households rely on a new generation of more-efficient biomass cookstoves. United Nations' Secretary-General Ban Ki-moon has said that renewable energy has the ability to lift the poorest nations to new levels of prosperity. At the national level, at least 30 nations around the world already have renewable energy contributing more than 20% of energy supply. National renewable energy markets are projected to continue to grow strongly in the coming decade and beyond, and some 120 countries have various policy targets for longer-term shares of renewable energy, including a 20% target of all electricity generated for the European Union by 2020. Some countries have much higher long-term policy targets of up to 100% renewables. Outside Europe, a diverse group of 20 or more other countries target renewable energy shares in the 2020–2030 time frame that range from 10% to 50%.

Renewable energy often displaces conventional fuels in four areas: electricity generation, hot water/space heating, transportation, and rural (off-grid) energy services:

Power Generation

Renewable hydroelectric energy provides 16.3% of the worlds electricity. When hydroelectric is combined with other renewables such as wind, geothermal, solar, biomass and waste: together they make the "renewables" total, 21.7% of electricity generation worldwide as of 2013. Renewable power generators are spread across many countries, and wind power alone already provides a significant share of electricity in some areas: for example, 14% in the U.S. state of Iowa, 40% in the northern German state of Schleswig-Holstein, and 49% in Denmark. Some countries get most of their power from renewables, including Iceland (100%), Norway (98%), Brazil (86%), Austria (62%), New Zealand (65%), and Sweden (54%).

Heating

Solar water heating makes an important contribution to renewable heat in many countries, most notably in China, which now has 70% of the global total (180 GWth). Most of these systems are installed on multi-family apartment buildings and meet a portion of the hot water needs of an estimated 50–60 million households in China. Worldwide, total installed solar water heating systems meet a portion of the water heating needs of over 70 million households. The use of biomass for heating continues to grow as well. In Sweden, national use of biomass energy has surpassed that of oil. Direct geothermal for heating is also growing rapidly. The newest addition to Heating is from Geothermal Heat Pumps which provide both heating and cooling, and also flatten the electric demand curve and are thus an increasing national priority.

Transportation

A bus fueled by biodiesel

Bioethanol is an alcohol made by fermentation, mostly from carbohydrates produced in sugar or starch crops such as corn, sugarcane, or sweet sorghum. Cellulosic biomass, derived from non-food sources such as trees and grasses is also being developed as a feedstock for ethanol production. Ethanol can be used as a fuel for vehicles in its pure form, but it is usually used as a gasoline additive to increase octane and improve vehicle emissions. Bioethanol is widely used in the USA and in Brazil. Biodiesel can be used as a fuel for vehicles in its pure form, but it is usually used as a diesel additive to reduce levels of particulates, carbon monoxide, and hydrocarbons from diesel-powered vehicles. Biodiesel is produced from oils or fats using transesterification and is the most common biofuel in Europe.

A solar vehicle is an electric vehicle powered completely or significantly by direct solar energy. Usually, photovoltaic (PV) cells contained in solar panels convert the sun's energy directly into electric energy. The term "solar vehicle" usually implies that solar energy is used to power all or part of a vehicle's propulsion. Solar power may be also used to provide power for communications or controls or other auxiliary functions. Solar powered boats have mainly been limited to rivers and canals, but in 2007 an experimental 14m catamaran, the Sun21 sailed the Atlantic from Seville to Miami, and from there to New York. It was the first crossing of the Atlantic powered only by solar. Solar vehicles are not sold as practical day-to-day transportation devices at present, but are primarily demonstration vehicles and engineering exercises, often sponsored by government agencies. However, indirectly solar-charged vehicles are widespread and solar boats are available commercially.

History

Prior to the development of coal in the mid 19th century, nearly all energy used was renewable. Almost without a doubt the oldest known use of renewable energy, in the form of traditional biomass to fuel fires, dates from 790,000 years ago. Use of biomass for fire did not become commonplace until many hundreds of thousands of years later, sometime between 200,000 and 400,000 years ago. Probably the second oldest usage of renewable energy is harnessing the wind in order to drive ships over water. This practice can be traced back some 7000 years, to ships on the Nile. Moving into the time of recorded history, the primary sources of traditional renewable energy were human labor, animal power, water power, wind, in grain crushing windmills, and firewood, a traditional biomass. A graph of energy use in the United States up until 1900 shows oil and natural gas with about the same importance in 1900 as wind and solar played in 2010.

In the 1860s and '70s there were already fears that civilization would run out of fossil fuels and the need was felt for a better source. In 1873 Professor Augustine Mouchot wrote:

The time will arrive when the industry of Europe will cease to find those natural resources, so necessary for it. Petroleum springs and coal mines are not inexhaustible but are rapidly diminishing in many places. Will man, then, return to the power of water and wind? Or will he emigrate where the most powerful source of heat sends its rays to all? History will show what will come.

In 1885, Werner von Siemens, commenting on the discovery of the photovoltaic effect in the solid state, wrote:

In conclusion, I would say that however great the scientific importance of this discovery may be, its practical value will be no less obvious when we reflect that the supply of solar energy is both without limit and without cost, and that it will continue to pour down upon us for countless ages after all the coal deposits of the earth have been exhausted and forgotten.

Max Weber mentioned the end of fossil fuel in the concluding paragraphs of his Die protestantische Ethik und der Geist des Kapitalismus, published in 1905.

Development of solar engines continued until the outbreak of World War I. The importance of solar energy was recognized in a 1911 *Scientific American* article: "in the far distant future, natural fuels having been exhausted [solar power] will remain as the only means of existence of the human race".

The theory of peak oil was published in 1956. In the 1970s environmentalists promoted the development of renewable energy both as a replacement for the eventual depletion of oil, as well as for an escape from dependence on oil, and the first electricity generating wind turbines appeared. Solar had long been used for heating and cooling, but solar panels were too costly to build solar farms until 1980.

The IEA 2014 World Energy Outlook projects a growth of renewable energy supply from 1,700 gigawatts in 2014 to 4,550 gigawatts in 2040. Fossil fuels received about $550 billion in subsidies in 2013, compared to $120 billion for all renewable energies.

Mainstream Technologies

Wind Power

The 845 MW Shepherds Flat Wind Farm near Arlington, Oregon, USA

Airflows can be used to run wind turbines. Modern utility-scale wind turbines range from around 600 kW to 5 MW of rated power, although turbines with rated output of 1.5–3 MW have become the most common for commercial use; the power available from the wind is a function of the cube of the wind speed, so as wind speed increases, power output increases up to the maximum output for the particular turbine. Areas where winds are stronger and more constant, such as offshore and high altitude sites, are preferred locations for wind farms. Typically full load hours of wind turbines vary between 16 and 57 percent annually, but might be higher in particularly favorable offshore sites.

Globally, the long-term technical potential of wind energy is believed to be five times total current global energy production, or 40 times current electricity demand, assuming all practical barriers needed were overcome. This would require wind turbines to be installed over large areas, particularly in areas of higher wind resources, such as offshore. As offshore wind speeds average ~90% greater than that of land, so offshore resources can contribute substantially more energy than land stationed turbines. In 2014 global wind generation was 706 terawatt-hours or 3% of the worlds total electricity.

Hydropower

The Three Gorges Dam on the Yangtze River in China

In 2015 hydropower generated 16.6% of the worlds total electricity and 70% of all renewable electricity. Since water is about 800 times denser than air, even a slow flowing stream of water, or moderate sea swell, can yield considerable amounts of energy. There are many forms of water energy:

- Historically hydroelectric power came from constructing large hydroelectric dams and reservoirs, which are still popular in third world countries. The largest of which is the Three Gorges Dam(2003) in China and the Itaipu Dam(1984) built by Brazil and Paraguay.

- Small hydro systems are hydroelectric power installations that typically produce up to 50 MW of power. They are often used on small rivers or as a low impact development on larger rivers. China is the largest producer of hydroelectricity in the world and has more than 45,000 small hydro installations.

- Run-of-the-river hydroelectricity plants derive kinetic energy from rivers without the creation of a large reservoir. This style of generation may still produce a large amount of electricity, such as the Chief Joseph Dam on the Columbia river in the United States.

Hydropower is produced in 150 countries, with the Asia-Pacific region generating 32 percent of global hydropower in 2010. For counties having the largest percentage of electricity from renewables, the top 50 are primarily hydroelectric. China is the largest hydroelectricity producer, with 721 terawatt-hours of production in 2010, representing around 17 percent of domestic electricity use. There are now three hydroelectricity stations larger than 10 GW: the Three Gorges Dam in China, Itaipu Dam across the Brazil/Paraguay border, and Guri Dam in Venezuela.

Wave power, which captures the energy of ocean surface waves, and tidal power, converting the energy of tides, are two forms of hydropower with future potential; however, they are not yet widely employed commercially. A demonstration project operated by the Ocean Renewable Power Company on the coast of Maine, and connected to the grid, harnesses tidal power from the Bay of Fundy, location of world's highest tidal flow. Ocean thermal energy conversion, which uses the temperature difference between cooler deep and warmer surface waters, has currently no economic feasibility.

Solar Energy

Satellite image of the 550-megawatt Topaz Solar Farm in California, USA

Solar energy, radiant light and heat from the sun, is harnessed using a range of ever-evolving technologies such as solar heating, photovoltaics, concentrated solar power (CSP), concentrator photovoltaics (CPV), solar architecture and artificial photosynthesis. Solar technologies are broadly characterized as either passive solar or active solar depending on the way they capture, convert and distribute solar energy. Passive solar techniques include orienting a building to the Sun, selecting materials with favorable thermal mass or light dispersing properties, and designing spaces that naturally circulate air. Active solar technologies encompass solar thermal energy, using solar collectors for heating, and solar power, converting sunlight into electricity either directly using photovoltaics (PV), or indirectly using concentrated solar power (CSP).

A photovoltaic system converts light into electrical direct current (DC) by taking advantage of the photoelectric effect. Solar PV has turned into a multi-billion, fast-growing industry, continues to improve its cost-effectiveness, and has the most potential of any renewable technologies together with CSP. Concentrated solar power (CSP) systems use lenses or mirrors and tracking systems to focus a large area of sunlight into a small beam. Commercial concentrated solar power plants were first developed in the 1980s. CSP-Stirling has by far the highest efficiency among all solar energy technologies.

In 2011, the International Energy Agency said that "the development of affordable, inexhaustible and clean solar energy technologies will have huge longer-term benefits. It will increase countries' energy security through reliance on an indigenous, inexhaustible and mostly import-independent resource, enhance sustainability, reduce pollution, lower the costs of mitigating climate change, and keep fossil fuel prices lower than otherwise. These advantages are global. Hence the additional costs of the incentives for early deployment should be considered learning investments; they must be wisely spent and need to be widely shared". In 2014 global solar generation was 186 terawatt-hours, slightly less than 1% of the worlds total grid electricity.

Geothermal Energy

Steam rising from the Nesjavellir Geothermal Power Station in Iceland

High Temperature Geothermal energy is from thermal energy generated and stored in the Earth. Thermal energy is the energy that determines the temperature of matter. Earth's geo-thermal energy originates from the original formation of the planet and from radioactive de-cay of minerals (in currently uncertain but possibly roughly equal proportions). The geother-mal gradient, which is the difference in temperature between the core of the planet and its surface, drives a continuous conduction of thermal energy in the form of heat from the core to the surface.

The heat that is used for geothermal energy can be from deep within the Earth, all the way down to Earth's core – 4,000 miles (6,400 km) down. At the core, temperatures may reach over 9,000 °F (5,000 °C). Heat conducts from the core to surrounding rock. Extremely high temperature and pressure cause some rock to melt, which is commonly known as magma. Magma convects upward since it is lighter than the solid rock. This magma then heats rock and water in the crust, some-times up to 700 °F (371 °C).

From hot springs, geothermal energy has been used for bathing since Paleolithic times and for space heating since ancient Roman times, but it is now better known for electricity generation.

Low Temperature Geothermal refers to the use of the outer crust of the earth as a Thermal Battery to facilitate Renewable thermal energy for heating and cooling buildings, and other refrigeration and industrial uses. In this form of Geothermal, a Geothermal Heat Pump and Ground-coupled heat exchanger are used together to move heat energy into the earth (for cooling) and out of the earth (for heating) on a varying seasonal basis. Low temperature Geothermal (generally referred to as "GHP") is an increasingly important renewable technology because it both reduces total annual energy loads associated with heating and cooling, and it also flattens the electric demand curve eliminating the extreme summer and winter peak electric supply requirements. Thus Low Temperature Geothermal/GHP is becoming an increasing national priority with multiple tax credit support and focus as part of the ongoing movement toward Net Zero Energy. New York City has even just passed a law to require GHP anytime is shown to be economical with 20 year financing including the Socialized Cost of Carbon.

Bio Energy

Sugarcane plantation to produce ethanol in Brazil

Biomass is biological material derived from living, or recently living organisms. It most often refers to plants or plant-derived materials which are specifically called lignocellulosic biomass. As an energy source, biomass can either be used directly via combustion to produce heat, or indirectly after converting it to various forms of biofuel. Conversion of biomass to biofuel can be achieved by different methods which are broadly classified into: *thermal*, *chemical*, and *biochemical* methods. Wood remains the largest biomass energy source today; examples include forest residues – such as dead trees, branches and tree stumps –, yard clippings, wood chips and even municipal solid waste. In the second sense, biomass includes plant or animal matter that can be converted into fibers or other industrial chemicals, including biofuels. Industrial biomass can be grown from numerous types of plants, including miscanthus, switchgrass, hemp, corn, poplar, willow, sorghum, sugarcane, bamboo, and a variety of tree species, ranging from eucalyptus to oil palm (palm oil).

Plant energy is produced by crops specifically grown for use as fuel that offer high biomass output per hectare with low input energy. Some examples of these plants are wheat, which typically yield 7.5–8 tonnes of grain per hectare, and straw, which typically yield 3.5–5 tonnes per hectare in the UK. The grain can be used for liquid transportation fuels while the straw can be burned to produce heat or electricity. Plant biomass can also be degraded from cellulose to glucose through a series of chemical treatments, and the resulting sugar can then be used as a first generation biofuel.

A CHP power station using wood to supply 30,000 households in France

Biomass can be converted to other usable forms of energy like methane gas or transportation fuels like ethanol and biodiesel. Rotting garbage, and agricultural and human waste, all release methane gas – also called landfill gas or biogas. Crops, such as corn and sugarcane, can be fermented to produce the transportation fuel, ethanol. Biodiesel, another transportation fuel, can be produced from left-over food products like vegetable oils and animal fats. Also, biomass to liquids (BTLs) and cellulosic ethanol are still under research. There is a great deal of research involving algal fuel or algae-derived biomass due to the fact that it's a non-food resource and can be produced at rates 5 to 10 times those of other types of land-based agriculture, such as corn and soy. Once harvested, it can be fermented to produce biofuels such as ethanol, butanol, and methane, as well as biodiesel and hydrogen. The biomass used for electricity generation varies by region. Forest by-products, such as wood residues, are common in the United States. Agricultural waste is common in Mauritius (sugar cane residue) and Southeast Asia (rice husks). Animal husbandry residues, such as poultry litter, are common in the United Kingdom.

Biofuels include a wide range of fuels which are derived from biomass. The term covers solid, liquid, and gaseous fuels. Liquid biofuels include bioalcohols, such as bioethanol, and oils, such as biodiesel. Gaseous biofuels include biogas, landfill gas and synthetic gas. Bioethanol is an alcohol made by fermenting the sugar components of plant materials and it is made mostly from sugar and starch crops. These include maize, sugarcane and, more recently, sweet sorghum. The latter crop is particularly suitable for growing in dryland conditions, and is being investigated by International Crops Research Institute for the Semi-Arid Tropics for its potential to provide fuel, along with food and animal feed, in arid parts of Asia and Africa.

With advanced technology being developed, cellulosic biomass, such as trees and grasses, are also used as feedstocks for ethanol production. Ethanol can be used as a fuel for vehicles in its pure form, but it is usually used as a gasoline additive to increase octane and improve vehicle emissions. Bioethanol is widely used in the United States and in Brazil. The energy costs for producing bio-ethanol are almost equal to, the energy yields from bio-ethanol. However, according to the European Environment Agency, biofuels do not address global warming concerns. Biodiesel is made from vegetable oils, animal fats or recycled greases. It can be used as a fuel for vehicles in its pure form, or more commonly as a diesel additive to reduce levels of particulates, carbon monoxide, and hydrocarbons from diesel-powered vehicles. Biodiesel is produced from oils or fats using transesterification and is the most common biofuel in Europe. Biofuels provided 2.7% of the world's transport fuel in 2010.

Biomass, biogas and biofuels are burned to produce heat/power and in doing so harm the environment. Pollutants such as sulphurous oxides (SO_x), nitrous oxides (NO_x), and particulate matter (PM) are produced from the combustion of biomass; the World Health Organisation estimates that 7 million premature deaths are caused each year by air pollution. Biomass combustion is a major contributor. The life cycle of the plants is sustainable, the lives of people less so.

Energy Storage

Energy storage is a collection of methods used to store electrical energy on an electrical power grid, or off it. Electrical energy is stored during times when production (especially from intermittent power plants such as renewable electricity sources such as wind power, tidal power, solar power) exceeds consumption, and returned to the grid when production falls below consumption. Water pumped into a hydroelectric dam is the largest form of power storage.

Market and Industry Trends

Growth of Renewables

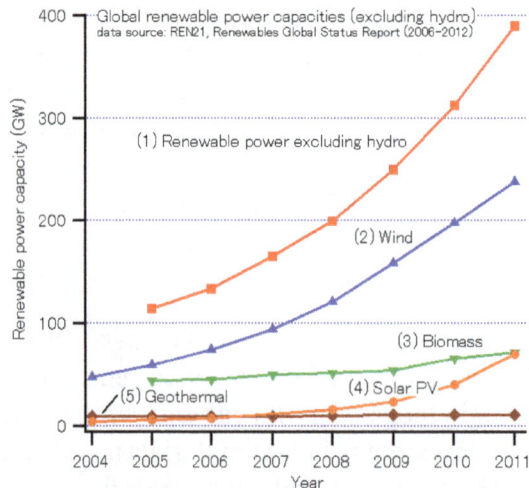

Global growth of renewables through to 2011

From the end of 2004, worldwide renewable energy capacity grew at rates of 10–60% annually for many technologies. For wind power and many other renewable technologies, growth accelerated in 2009 relative to the previous four years. More wind power capacity was added during 2009 than any other renewable technology. However, grid-connected PV increased the fastest of all renewables technologies, with a 60% annual average growth rate. In 2010, renewable power constituted about a third of the newly built power generation capacities.

Projections vary, but scientists have advanced a plan to power 100% of the world's energy with wind, hydroelectric, and solar power by the year 2030.

According to a 2011 projection by the International Energy Agency, solar power generators may produce most of the world's electricity within 50 years, reducing the emissions of greenhouse gases that harm the environment. Cedric Philibert, senior analyst in the renewable energy division at the IEA said: "Photovoltaic and solar-thermal plants may meet most of the world's demand for electricity by 2060 – and half of all energy needs – with wind, hydropower and biomass plants supplying much of the remaining generation". "Photovoltaic and concentrated solar power together can become the major source of electricity", Philibert said.

In 2014 global wind power capacity expanded 16% to 369,553 MW. Yearly wind energy production is also growing rapidly and has reached around 4% of worldwide electricity usage, 11.4% in the EU, and it is widely used in Asia, and the United States. In 2015, worldwide installed photovoltaics capacity increased to 227 gigawatts (GW), sufficient to supply 1 percent of global electricity demands. Solar thermal energy stations operate in the USA and Spain, and as of 2016, the largest of these is the 392 MW Ivanpah Solar Electric Generating System in California. The world's largest geothermal power installation is The Geysers in California, with a rated capacity of 750 MW. Brazil has one of the largest renewable energy programs in the world, involving production of ethanol fuel from sugar cane, and ethanol now provides 18% of the country's automotive fuel. Ethanol fuel is also widely available in the USA.

Selected renewable energy global indicators	2008	2009	2010	2011	2012	2013	2014	2015
Investment in new renewable capacity (annual) (10^9 USD)	182	178	237	279	256	232	270	285
Renewables power capacity (existing) (GWe)	1,140	1,230	1,320	1,360	1,470	1,578	1,712	1,849
Hydropower capacity (existing) (GWe)	885	915	945	970	990	1,018	1,055	1,064
Wind power capacity (existing) (GWe)	121	159	198	238	283	319	370	433
Solar PV capacity (grid-connected) (GWe)	16	23	40	70	100	138	177	227
Solar hot water capacity (existing) (GWth)	130	160	185	232	255	373	406	435
Ethanol production (annual) (10^9 litres)	67	76	86	86	83	87	94	98
Biodiesel production (annual) (10^9 litres)	12	17.8	18.5	21.4	22.5	26	29.7	30
Countries with policy targets for renewable energy use	79	89	98	118	138	144	164	173

Source: The Renewable Energy Policy Network for the 21st Century (REN21)–Global Status Report

Economic Trends

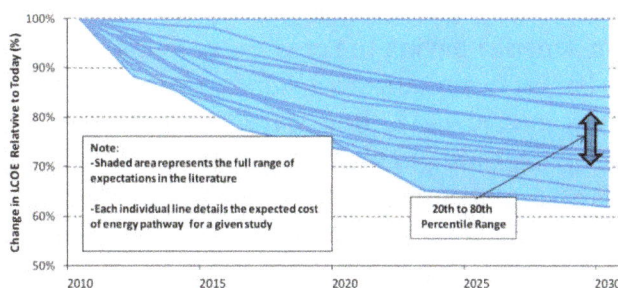

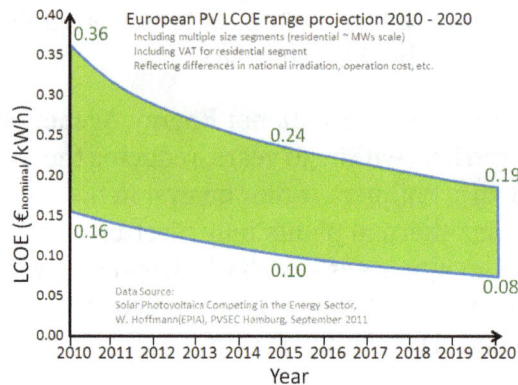

Projection of levelized cost for wind in the U.S. (left) and solar power in Europe

Renewable energy technologies are getting cheaper, through technological change and through the benefits of mass production and market competition. A 2011 IEA report said: "A portfolio of renewable energy technologies is becoming cost-competitive in an increasingly broad range of circumstances, in some cases providing investment opportunities without the need for specific economic support," and added that "cost reductions in critical technologies, such as wind and solar, are set to continue."

Hydro-electricity and geothermal electricity produced at favourable sites are now the cheapest way to generate electricity. Renewable energy costs continue to drop, and the levelised cost of electricity (LCOE) is declining for wind power, solar photovoltaic (PV), concentrated solar power (CSP) and some biomass technologies. Renewable energy is also the most economic solution for new grid-connected capacity in areas with good resources. As the cost of renewable power falls, the scope of economically viable applications increases. Renewable technologies are now often the most economic solution for new generating capacity. Where "oil-fired generation is the predominant power generation source (e.g. on islands, off-grid and in some countries) a lower-cost renewable solution almost always exists today". A series of studies by the US National Renewable Energy Laboratory modeled the "grid in the Western US under a number of different scenarios where intermittent renewables accounted for 33 percent of the total power." In the models, inefficiencies in cycling the fossil fuel plants to compensate for the variation in solar and wind energy resulted in an additional cost of "between $0.47 and $1.28 to each MegaWatt hour generated"; however, the savings in the cost of the fuels saved "adds up to $7 billion, meaning the added costs are, at most, two percent of the savings."

Hydroelectricity

Only 25% of the worlds estimated hydroelectric potential of 14,000 TWh/year has been developed, with Africa, Asia and Latin America having the greatest potential. The Three Gorges Dam in Hubei, China, has the world's largest instantaneous generating capacity (22,500 MW), with the Itaipu Dam in Brazil/Paraguay in second place (14,000 MW). The Three Gorges Dam is operated jointly with the much smaller Gezhouba Dam (3,115 MW). As of 2012[update], the total generating capacity of this two-dam complex is 25,615 MW. In 2008, this complex generated 98 TWh of electricity (81 TWh from the Three Gorges Dam and 17 TWh from the Gezhouba Dam), which is 3% more power in one year than the 95 TWh generated by Itaipu in 2008.

Wind Power Development

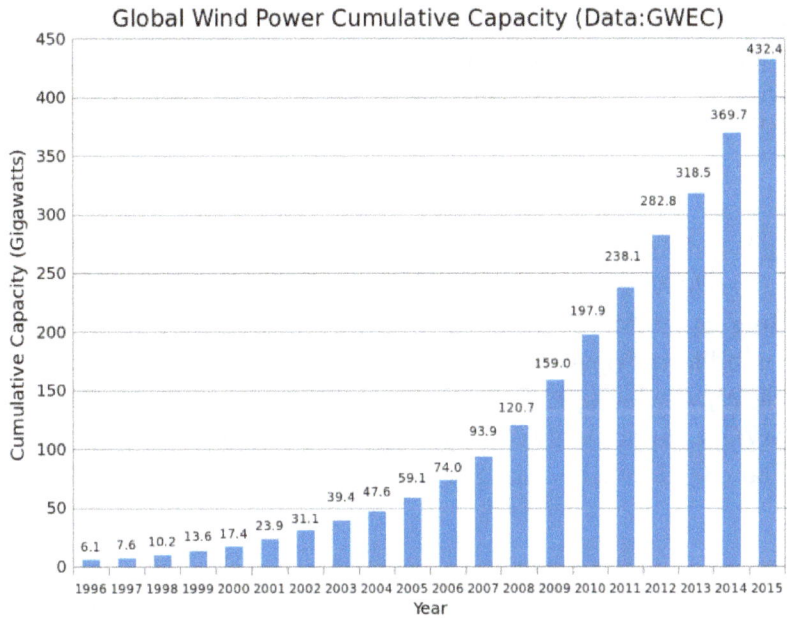

Worldwide growth of wind capacity (1996–2014)

Four offshore wind farms are in the Thames Estuary area: Kentish Flats, Gunfleet Sands, Thanet and London Array. The latter is the largest in the world as of April 2013.

Wind power is widely used in Europe, China, and the United States. From 2004 to 2014, worldwide installed capacity of wind power has been growing from 47 GW to 369 GW—a more than sevenfold increase within 10 years with 2014 breaking a new record in global installations (51 GW). As of the end of 2014, China, the United States and Germany combined accounted for half of total global capacity. Several other countries have achieved relatively high levels of wind power penetration, such as 21% of stationary electricity production in Denmark, 18% in Portugal, 16% in Spain, and 14% in Ireland in 2010 and have since continued to expand their installed capacity. More than 80 countries around the world are using wind power on a commercial basis.

Offshore Wind Power

As of 2014, offshore wind power amounted to 8,771 megawatt of global installed capacity. Although offshore capacity doubled within three years (from 4,117 MW in 2011), it accounted for only 2.3% of the total wind power capacity. The United Kingdom is the undisputed leader of offshore power with half of the world's installed capacity ahead of Denmark, Germany, Belgium and China.

List of Offshore and Onshore Wind Farms

As of 2012, the Alta Wind Energy Center (California, 1,020 MW) is the world's largest wind farm. The London Array (630 MW) is the largest offshore wind farm in the world. The United Kingdom is the world's leading generator of offshore wind power, followed by Denmark. There are several large offshore wind farms operational and under construction and these include Anholt (400 MW), BARD (400 MW), Clyde (548 MW), Fântânele-Cogealac (600 MW), Greater Gabbard (500 MW), Lincs (270 MW), London Array (630 MW), Lower Snake River (343 MW), Macarthur (420 MW), Shepherds Flat (845 MW), and the Sheringham Shoal (317 MW).

Solar Thermal

The 377 MW Ivanpah Solar Electric Generating System with all three towers under load, Feb 2014. Taken from I-15.

The United States conducted much early research in photovoltaics and concentrated solar power. The U.S. is among the top countries in the world in electricity generated by the Sun and several of the world's largest utility-scale installations are located in the desert Southwest.

Solar Towers of the PS10 and PS20 solar thermal plants in Spain

The oldest solar thermal power plant in the world is the 354 megawatt (MW) SEGS thermal power plant, in California. The Ivanpah Solar Electric Generating System is a solar thermal power project in the California Mojave Desert, 40 miles (64 km) southwest of Las Vegas, with a gross capacity of 377 MW. The 280 MW Solana Generating Station is a solar power plant near Gila Bend, Arizona, about 70 miles (110 km) southwest of Phoenix, completed in 2013. When commissioned it was the largest parabolic trough plant in the world and the first U.S. solar plant with molten salt thermal energy storage.

The solar thermal power industry is growing rapidly with 1.3 GW under construction in 2012 and more planned. Spain is the epicenter of solar thermal power development with 873 MW under construction, and a further 271 MW under development. In the United States, 5,600 MW of solar thermal power projects have been announced. Several power plants have been constructed in the Mojave Desert, Southwestern United States. The Ivanpah Solar Power Facility being the most recent. In developing countries, three World Bank projects for integrated solar thermal/combined-cycle gas-turbine power plants in Egypt, Mexico, and Morocco have been approved.

Photovoltaic Development

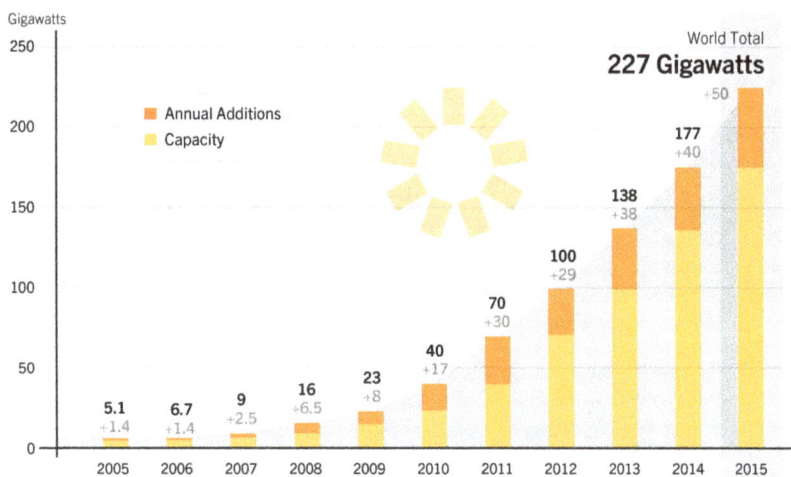

Worldwide growth of PV capacity grouped by region in MW (2006–2014)

Photovoltaics (PV) uses solar cells assembled into solar panels to convert sunlight into electricity. It's a fast-growing technology doubling its worldwide installed capacity every couple of years. PV systems range from small, residential and commercial rooftop or building integrated installations, to large utility-scale photovoltaic power station. The predominant PV technology is crystalline silicon, while thin-film solar cell technology accounts for about 10 percent of global photovoltaic deployment. In recent years, PV technology has improved its electricity generating efficiency, reduced the installation cost per watt as well as its energy payback time, and has reached grid parity in at least 30 different markets by 2014. Financial institutions are predicting a second solar "gold rush" in the near future.

At the end of 2014, worldwide PV capacity reached at least 177,000 megawatts. Photovoltaics grew fastest in China, followed by Japan and the United States, while Germany remains the world's largest overall producer of photovoltaic power, contributing about 7.0 percent to the overall electricity generation. Italy meets 7.9 percent of its electricity demands with photovoltaic power—the highest share worldwide. For 2015, global cumulative capacity is forecasted to increase by more than 50 gigawatts (GW). By 2018, worldwide capacity is projected to reach as much as 430 gigawatts. This corresponds to a tripling within five years. Solar power is forecasted to become the world's largest source of electricity by 2050, with solar photovoltaics and concentrated solar power contributing 16% and 11%, respectively. This requires an increase of installed PV capacity to 4,600 GW, of which more than half is expected to be deployed in China and India.

Photovoltaic Power Stations

Solar panels at the 550 MW Topaz Solar Farm

Nellis Solar Power Plant, photovoltaic power plant in Nevada, USA

Commercial concentrated solar power plants were first developed in the 1980s. As the cost of solar electricity has fallen, the number of grid-connected solar PV systems has grown into the millions and utility-scale solar power stations with hundreds of megawatts are being built. Solar PV is rapidly becoming an inexpensive, low-carbon technology to harness renewable energy from the Sun.

Many solar photovoltaic power stations have been built, mainly in Europe, China and the USA. The 579 MW Solar Star, in the United States, is the world's largest PV power station.

Many of these plants are integrated with agriculture and some use tracking systems that follow the sun's daily path across the sky to generate more electricity than fixed-mounted systems. There are no fuel costs or emissions during operation of the power stations.

However, when it comes to renewable energy systems and PV, it is not just large systems that matter. Building-integrated photovoltaics or "onsite" PV systems use existing land and structures and generate power close to where it is consumed.

Biofuel Development

Brazil produces bioethanol made from sugarcane available throughout the country. A typical gas station with dual fuel service is marked "A" for alcohol (ethanol) and "G" for gasoline.

Biofuels provided 3% of the world's transport fuel in 2010. Mandates for blending biofuels exist in 31 countries at the national level and in 29 states/provinces. According to the International Energy Agency, biofuels have the potential to meet more than a quarter of world demand for transportation fuels by 2050.

Since the 1970s, Brazil has had an ethanol fuel program which has allowed the country to become the world's second largest producer of ethanol (after the United States) and the world's largest exporter. Brazil's ethanol fuel program uses modern equipment and cheap sugarcane as feedstock, and the residual cane-waste (bagasse) is used to produce heat and power. There are no longer light vehicles in Brazil running on pure gasoline. By the end of 2008 there were 35,000 filling stations throughout Brazil with at least one ethanol pump. Unfortunately, Operation Car Wash has seriously eroded public trust in oil companies and has implicated several high ranking Brazilian officials.

Nearly all the gasoline sold in the United States today is mixed with 10% ethanol, and motor vehicle manufacturers already produce vehicles designed to run on much higher ethanol blends. Ford, Daimler AG, and GM are among the automobile companies that sell "flexible-fuel" cars, trucks, and minivans that can use gasoline and ethanol blends ranging from pure gasoline up to 85% ethanol. By mid-2006, there were approximately 6 million ethanol compatible vehicles on U.S. roads.

Geothermal Development

Geothermal plant at The Geysers, California, USA

Geothermal power is cost effective, reliable, sustainable, and environmentally friendly, but has historically been limited to areas near tectonic plate boundaries. Recent technological advances have expanded the range and size of viable resources, especially for applications such as home heating, opening a potential for widespread exploitation. Geothermal wells release greenhouse gases trapped deep within the earth, but these emissions are much lower per energy unit than those of fossil fuels. As a result, geothermal power has the potential to help mitigate global warming if widely deployed in place of fossil fuels.

The International Geothermal Association (IGA) has reported that 10,715 MW of geothermal power in 24 countries is online, which is expected to generate 67,246 GWh of electricity in 2010. This represents a 20% increase in geothermal power online capacity since 2005. IGA projects this will grow to 18,500 MW by 2015, due to the large number of projects presently under consideration, often in areas previously assumed to have little exploitable resource.

In 2010, the United States led the world in geothermal electricity production with 3,086 MW of installed capacity from 77 power plants; the largest group of geothermal power plants in the world is located at The Geysers, a geothermal field in California. The Philippines follows the US as the second highest producer of geothermal power in the world, with 1,904 MW of capacity online; geothermal power makes up approximately 18% of the country's electricity generation.

Developing Countries

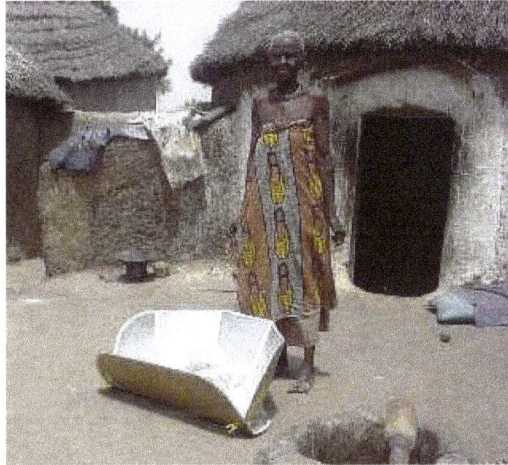

Solar cookers use sunlight as energy source for outdoor cooking

Renewable energy technology has sometimes been seen as a costly luxury item by critics, and affordable only in the affluent developed world. This erroneous view has persisted for many years, but 2015 was the first year when investment in non-hydro renewables, was higher in developing countries, with $156 billion invested, mainly in China, India, and Brazil.

Renewable energy can be particularly suitable for developing countries. In rural and remote areas, transmission and distribution of energy generated from fossil fuels can be difficult and expensive. Producing renewable energy locally can offer a viable alternative.

Technology advances are opening up a huge new market for solar power: the approximately 1.3 billion people around the world who don't have access to grid electricity. Even though they are typically very poor, these people have to pay far more for lighting than people in rich countries because they use inefficient kerosene lamps. Solar power costs half as much as lighting with kerosene. As of 2010, an estimated 3 million households get power from small solar PV systems. Kenya is the world leader in the number of solar power systems installed per capita. More than 30,000 very small solar panels, each producing 12 to 30 watts, are sold in Kenya annually. Some Small Island Developing States (SIDS) are also turning to solar power to reduce their costs and increase their sustainability.

Micro-hydro configured into mini-grids also provide power. Over 44 million households use biogas made in household-scale digesters for lighting and/or cooking, and more than 166 million households rely on a new generation of more-efficient biomass cookstoves. Clean liquid fuel sourced from renewable feedstocks are used for cooking and lighting in energy-poor areas of the developing world. Alcohol fuels (ethanol and methanol) can be produced sustainably from non-food sugary, starchy, and cellulostic feedstocks. Project Gaia, Inc. and CleanStar Mozambique are implementing clean cooking programs with liquid ethanol stoves in Ethiopia, Kenya, Nigeria and Mozambique.

Renewable energy projects in many developing countries have demonstrated that renewable energy can directly contribute to poverty reduction by providing the energy needed for creating businesses and employment. Renewable energy technologies can also make indirect contributions to alleviating poverty by providing energy for cooking, space heating, and lighting. Renewable energy can also contribute to education, by providing electricity to schools.

Industry and Policy Trends

Global new investment in renewable energy

data source: Bloomberg New Energy Finance, UNEP SEFI, Frankfurt School,
Global Trends in Renewable Energy Investment 2011

□ Small Distributed Capacity, RD&D

■ Financial New Investment

(Billion USD)

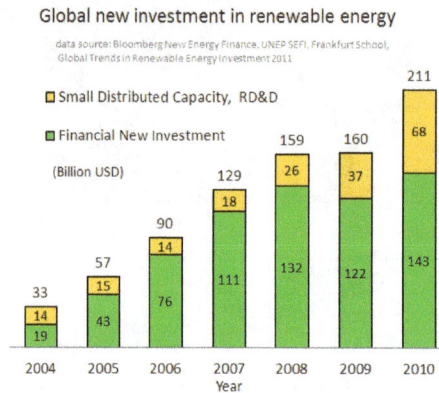

Global New Investments in Renewable Energy

U.S. President Barack Obama's American Recovery and Reinvestment Act of 2009 includes more than $70 billion in direct spending and tax credits for clean energy and associated transportation programs. Leading renewable energy companies include First Solar, Gamesa, GE Energy, Hanwha Q Cells, Sharp Solar, Siemens, SunOpta, Suntech Power, and Vestas.

Many national, state, and local governments have also created green banks. A green bank is a quasi-public financial institution that uses public capital to leverage private investment in clean energy technologies. Green banks use a variety of financial tools to bridge market gaps that hinder the deployment of clean energy.

The military has also focused on the use of renewable fuels for military vehicles. Unlike fossil fuels, renewable fuels can be produced in any country, creating a strategic advantage. The US military has already committed itself to have 50% of its energy consumption come from alternative sources.

The International Renewable Energy Agency (IRENA) is an intergovernmental organization for promoting the adoption of renewable energy worldwide. It aims to provide concrete policy advice and facilitate capacity building and technology transfer. IRENA was formed on 26 January 2009, by 75 countries signing the charter of IRENA. As of March 2010, IRENA has 143 member states who all are considered as founding members, of which 14 have also ratified the statute.

As of 2011, 119 countries have some form of national renewable energy policy target or renewable support policy. National targets now exist in at least 98 countries. There is also a wide range of policies at state/provincial and local levels.

United Nations' Secretary-General Ban Ki-moon has said that renewable energy has the ability to lift the poorest nations to new levels of prosperity. In October 2011, he "announced the creation of a high-level group to drum up support for energy access, energy efficiency and greater use of renewable energy. The group is to be co-chaired by Kandeh Yumkella, the chair of UN Energy and director general of the UN Industrial Development Organisation, and Charles Holliday, chairman of Bank of America".

100% Renewable Energy

The incentive to use 100% renewable energy, for electricity, transport, or even total primary energy supply globally, has been motivated by global warming and other ecological as well as economic concerns. The Intergovernmental Panel on Climate Change has said that there are few fundamental technological limits to integrating a portfolio of renewable energy technologies to meet most of total global energy demand. Renewable energy use has grown much faster than even advocates anticipated. At the national level, at least 30 nations around the world already have renewable energy contributing more than 20% of energy supply. Also, Professors S. Pacala and Robert H. Socolow have developed a series of "stabilization wedges" that can allow us to maintain our quality of life while avoiding catastrophic climate change, and "renewable energy sources," in aggregate, constitute the largest number of their "wedges."

Using 100% renewable energy was first suggested in a Science paper published in 1975 by Danish physicist Bent Sørensen. It was followed by several other proposals, until in 1998 the first detailed analysis of scenarios with very high shares of renewables were published. These were followed by the first detailed 100% scenarios. In 2006 a PhD thesis was published by Czisch in which it was shown that in a 100% renewable scenario energy supply could match demand in every hour of the year in Europa and North Africa. In the same year Danish Energy professor Henrik Lund published a first paper in which he addresses the optimal combination of renewables, which was followed by several other papers on the transition to 100% renewable energy in Denmark. Since then Lund has been publishing several papers on 100% renewable energy. After 2009 publications began to rise steeply, covering 100% scenarios for countries in Europa, America, Australia and other parts of the world.

In 2011 Mark Z. Jacobson, professor of civil and environmental engineering at Stanford University, and Mark Delucchi published a study on 100% renewable global energy supply in the journal Energy Policy. They found producing all new energy with wind power, solar power, and hydropower by 2030 is feasible and existing energy supply arrangements could be replaced by 2050. Barriers to implementing the renewable energy plan are seen to be "primarily social and political, not technological or economic". They also found that energy costs with a wind, solar, water system should be similar to today's energy costs.

Similarly, in the United States, the independent National Research Council has noted that "sufficient domestic renewable resources exist to allow renewable electricity to play a significant role in future electricity generation and thus help confront issues related to climate change, energy security, and the escalation of energy costs ... Renewable energy is an attractive option because renewable resources available in the United States, taken collectively, can supply significantly greater amounts of electricity than the total current or projected domestic demand." .

The most significant barriers to the widespread implementation of large-scale renewable energy and low carbon energy strategies are primarily political and not technological. According to the 2013 *Post Carbon Pathways* report, which reviewed many international studies, the key roadblocks are: climate change denial, the fossil fuels lobby, political inaction, unsustainable energy consumption, outdated energy infrastructure, and financial constraints.

Emerging Technologies

Other renewable energy technologies are still under development, and include cellulosic ethanol, hot-dry-rock geothermal power, and marine energy. These technologies are not yet widely demonstrated or have limited commercialization. Many are on the horizon and may have potential comparable to other renewable energy technologies, but still depend on attracting sufficient attention and research, development and demonstration (RD&D) funding.

There are numerous organizations within the academic, federal, and commercial sectors conducting large scale advanced research in the field of renewable energy. This research spans several areas of focus across the renewable energy spectrum. Most of the research is targeted at improving efficiency and increasing overall energy yields. Multiple federally supported research organizations have focused on renewable energy in recent years. Two of the most prominent of these labs are Sandia National Laboratories and the National Renewable Energy Laboratory (NREL), both of which are funded by the United States Department of Energy and supported by various corporate partners. Sandia has a total budget of $2.4 billion while NREL has a budget of $375 million.

Enhanced Geothermal System

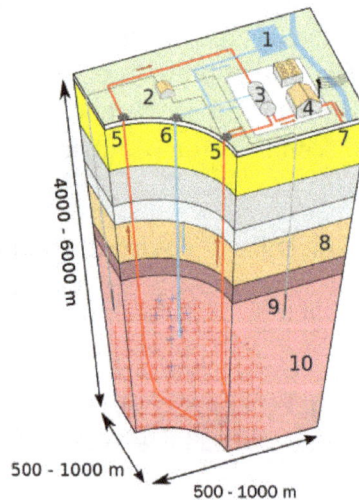

Enhanced geothermal system

Enhanced geothermal systems (EGS) are a new type of geothermal power technologies that do not require natural convective hydrothermal resources. The vast majority of geothermal energy within drilling reach is in dry and non-porous rock. EGS technologies "enhance" and/or create geothermal resources in this "hot dry rock (HDR)" through hydraulic stimulation. EGS and HDR technologies, like hydrothermal geothermal, are expected to be baseload resources which produce power 24 hours a day like a fossil plant. Distinct from hydrothermal, HDR and EGS may be feasible anywhere in the world, depending on the economic limits of drill depth. Good locations are over deep granite covered by a thick (3–5 km) layer of insulating sediments which slow heat loss. There are HDR and EGS systems currently being developed and tested in France, Australia, Japan, Germany, the U.S. and Switzerland. The largest EGS project in the world is a 25 megawatt demonstration plant currently being developed in the Cooper Basin, Australia. The Cooper Basin has the potential to generate 5,000–10,000 MW.

Cellulosic Ethanol

Several refineries that can process biomass and turn it into ethano are built by companies such as Iogen, POET, and Abengoa, while other companies such as the Verenium Corporation, Novozymes, and Dyadic International are producing enzymes which could enable future commercialization. The shift from food crop feedstocks to waste residues and native grasses offers significant opportunities for a range of players, from farmers to biotechnology firms, and from project developers to investors.

Marine Energy

Rance Tidal Power Station, France

Marine energy (also sometimes referred to as ocean energy) refers to the energy carried by ocean waves, tides, salinity, and ocean temperature differences. The movement of water in the world's oceans creates a vast store of kinetic energy, or energy in motion. This energy can be harnessed to generate electricity to power homes, transport and industries. The term marine energy encompasses both wave power – power from surface waves, and tidal power – obtained from the kinetic energy of large bodies of moving water. Reverse electrodialysis (RED) is a technology for generating electricity by mixing fresh river water and salty sea water in large power cells designed for this purpose; as of 2016 it is being tested at a small scale (50 kW). Offshore wind power is not a form of marine energy, as wind power is derived from the wind, even if the wind turbines are placed over water. The oceans have a tremendous amount of energy and are close to many if not most concentrated populations. Ocean energy has the potential of providing a substantial amount of new renewable energy around the world.

#	Station	Country	Location	Capacity
1.	Sihwa Lake Tidal Power Station	South Korea	37°18′47″N 126°36′46″E / 37.31306°N 126.61278°E / 37.31306; 126.61278 (Sihwa Lake Tidal Power Station)	254 MW
2.	Rance Tidal Power Station	France	48°37′05″N 02°01′24″W / 48.61806°N 2.02333°W / 48.61806; -2.02333 (Rance Tidal Power Station)	240 MW
3.	Annapolis Royal Generating Station	Canada	44°45′07″N 65°30′40″W / 44.75194°N 65.51111°W / 44.75194; -65.51111 (Annapolis Royal Generating Station)	20 MW

Experimental Solar Power

Concentrated photovoltaics (CPV) systems employ sunlight concentrated onto photovoltaic surfaces for the purpose of electricity generation. Thermoelectric, or "thermovoltaic" devices convert a temperature difference between dissimilar materials into an electric current.

Floating Solar Arrays

Floating solar arrays are PV systems that float on the surface of drinking water reservoirs, quarry lakes, irrigation canals or remediation and tailing ponds. A small number of such systems exist in France, India, Japan, South Korea, the United Kingdom, Singapore and the United States. The systems are said to have advantages over photovoltaics on land. The cost of land is more expensive, and there are fewer rules and regulations for structures built on bodies of water not used for recreation. Unlike most land-based solar plants, floating arrays can be unobtrusive because they are hidden from public view. They achieve higher efficiencies than PV panels on land, because water cools the panels. The panels have a special coating to prevent rust or corrosion. In May 2008, the Far Niente Winery in Oakville, California, pioneered the world's first floatovoltaic system by installing 994 solar PV modules with a total capacity of 477 kW onto 130 pontoons and floating them on the winery's irrigation pond. Utility-scale floating PV farms are starting to be built. Kyocera will develop the world's largest, a 13.4 MW farm on the reservoir above Yamakura Dam in Chiba Prefecture using 50,000 solar panels. Salt-water resistant floating farms are also being constructed for ocean use. The largest so far announced floatovoltaic project is a 350 MW power station in the Amazon region of Brazil.

Solar-assisted Heat Pump

A heat pump is a device that provides heat energy from a source of heat to a destination called a "heat sink". Heat pumps are designed to move thermal energy opposite to the direction of spontaneous heat flow by absorbing heat from a cold space and releasing it to a warmer one. A solar-assisted heat pump represents the integration of a heat pump and thermal solar panels in a single integrated system. Typically these two technologies are used separately (or only placing them in parallel) to produce hot water. In this system the solar thermal panel performs the function of the low temperature heat source and the heat produced is used to feed the heat pump's evaporator. The goal of this system is to get high COP and then produce energy in a more efficient and less expensive way.

It is possible to use any type of solar thermal panel (sheet and tubes, roll-bond, heat pipe, thermal plates) or hybrid (mono/polycrystalline, thin film) in combination with the heat pump. The use of a hybrid panel is preferable because it allows to cover a part of the electricity demand of the heat pump and reduce the power consumption and consequently the variable costs of the system.

Artificial Photosynthesis

Artificial photosynthesis uses techniques including nanotechnology to store solar electromagnetic energy in chemical bonds by splitting water to produce hydrogen and then using carbon dioxide to make methanol. Researchers in this field are striving to design molecular mimics of photosynthesis that utilize a wider region of the solar spectrum, employ catalytic systems made from abundant, inexpensive materials that are robust, readily repaired, non-toxic, stable in a variety of environmental conditions and perform more efficiently allowing a greater proportion of photon energy to end up in the storage compounds, i.e., carbohydrates (rather than building and sustaining living cells). However, prominent research faces hurdles, Sun Catalytix a MIT spin-off stopped scaling up their prototype fuel-cell in 2012, because it offers few savings over other ways to make hydrogen from sunlight.

Algae Fuels

Producing liquid fuels from oil-rich varieties of algae is an ongoing research topic. Various micro-algae grown in open or closed systems are being tried including some system that can be set up in brownfield and desert lands.

Solar Aircraft

In 2016, *Solar Impulse 2* was the first solar-powered aircraft to complete circumnavigation of the world.

An electric aircraft is an aircraft that runs on electric motors rather than internal combustion engines, with electricity coming from fuel cells, solar cells, ultracapacitors, power beaming, or batteries.

Currently, flying manned electric aircraft are mostly experimental demonstrators, though many small unmanned aerial vehicles are powered by batteries. Electrically powered model aircraft have been flown since the 1970s, with one report in 1957. The first man-carrying electrically powered flights were made in 1973. Between 2015-2016, a manned, solar-powered plane, Solar Impulse 2, completed a circumnavigation of the Earth.

Solar Updraft Tower

The Solar updraft tower is a renewable-energy power plant for generating electricity from low temperature solar heat. Sunshine heats the air beneath a very wide greenhouse-like roofed collector structure surrounding the central base of a very tall chimney tower. The resulting convection causes a hot air updraft in the tower by the chimney effect. This airflow drives wind turbines placed in the chimney updraft or around the chimney base to produce electricity. Plans for scaled-up versions of demonstration models will allow significant power generation, and may allow development of other applications, such as water extraction or distillation, and agriculture or horticulture.

A more advanced version of a similarly themed technology is the Vortex engine which aims to replace large physical chimneys with a vortex of air created by a shorter, less-expensive structure.

Space-based Solar Power

For either photovoltaic or thermal systems, one option is to loft them into space, particularly Geo-synchronous orbit. To be competitive with Earth-based solar power systems, the specific mass (kg/kW) times the cost to loft mass plus the cost of the parts needs to be $2400 or less. I.e., for a parts cost plus rectenna of $1100/kW, the product of the $/kg and kg/kW must be $1300/kW or less. Thus for 6.5 kg/kW, the transport cost cannot exceed $200/kg. While that will require a 100 to one reduction, SpaceX is targeting a ten to one reduction, Reaction Engines may make a 100 to one reduction possible.

Carbon-neutral and Negative Fuels

Carbon-neutral fuels are synthetic fuels (including methane, gasoline, diesel fuel, jet fuel or ammonia) produced by hydrogenating waste carbon dioxide recycled from power plant flue-gas emissions, recovered from automotive exhaust gas, or derived from carbonic acid in seawater. Such fuels are considered carbon-neutral because they do not result in a net increase in atmospheric greenhouse gases.

Carbon-neutral fuels offer relatively low cost energy storage, alleviating the problems of wind and solar variability, and they enable distribution of wind, water, and solar power through existing natural gas pipelines. Nighttime wind power is considered the most economical form of electrical power with which to synthesize fuel, because the load curve for electricity peaks sharply during the warmest hours of the day, but wind tends to blow slightly more at night than during the day, so, the price of nighttime wind power is often much less expensive than any alternative. Germany has built a 250 kilowatt synthetic methane plant which they are scaling up to 10 megawatts.

The George Olah carbon dioxide recycling plant in Grindavík, Iceland has been producing 2 million liters of methanol transportation fuel per year from flue exhaust of the Svartsengi Power Station since 2011. It has the capacity to produce 5 million liters per year.

Debate

Renewable electricity production, from sources such as wind power and solar power, is sometimes criticized for being variable or intermittent, but is not true for concentrated solar, geothermal and biofuels, that have continuity. In any case, the International Energy Agency has stated that deployment of renewable technologies usually increases the diversity of electricity sources and, through local generation, contributes to the flexibility of the system and its resistance to central shocks.

There have been "not in my back yard" (NIMBY) concerns relating to the visual and other impacts of some wind farms, with local residents sometimes fighting or blocking construction. In the USA, the Massachusetts Cape Wind project was delayed for years partly because of aesthetic concerns. However, residents in other areas have been more positive. According to a town councilor, the overwhelming majority of locals believe that the Ardrossan Wind Farm in Scotland has enhanced the area.

A recent UK Government document states that "projects are generally more likely to succeed if they have broad public support and the consent of local communities. This means giving communities both a say and a stake". In countries such as Germany and Denmark many renewable projects are owned by communities, particularly through cooperative structures, and contribute significantly to overall levels of renewable energy deployment.

The market for renewable energy technologies has continued to grow. Climate change concerns and increasing in green jobs, coupled with high oil prices, peak oil, oil wars, oil spills, promotion of electric vehicles and renewable electricity, nuclear disasters and increasing government support, are driving increasing renewable energy legislation, incentives and commercialization. New government spending, regulation and policies helped the industry weather the 2009 economic crisis better than many other sectors.

Gallery

Burbo, NW-England

Sunrise at the Fenton Wind Farm in Minnesota, USA

The CSP-station Andasol in Andalusia, Spain

Ivanpah solar plant in the Mojave Desert, California, United States

Three Gorges Dam and Gezhouba Dam, China

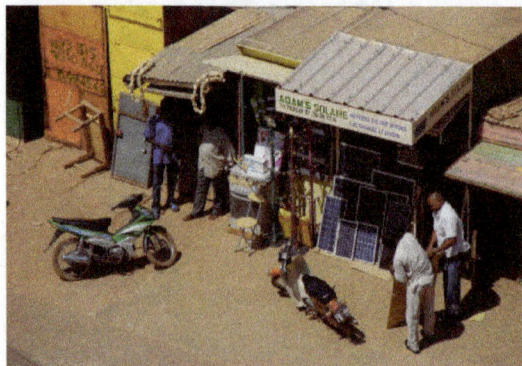

Shop selling PV panels in Ouagadougou, Burkina Faso

Stump harvesting increases recovery of biomass from forests

A small, roof-top mounted PV system in Bonn, Germany

The community-owned Westmill Solar Park in South East England

Komekurayama a photovoltaic power station in Kofu, Japan

Krafla, a geothermal power station in Iceland

Renewable Energy Resources:
Solar Power

A solar PV array on a rooftop in Hong Kong

The first three concentrated solar power (CSP) units of Spain's Solnova Solar Power Station in the foreground, with the PS10 and PS20 solar power towers in the background

Solar power is the conversion of sunlight into electricity, either directly using photovoltaics (PV), or indirectly using concentrated solar power (CSP). Concentrated solar power systems use lenses or mirrors and tracking systems to focus a large area of sunlight into a small beam. Photovoltaics convert light into an electric current using the photovoltaic effect.

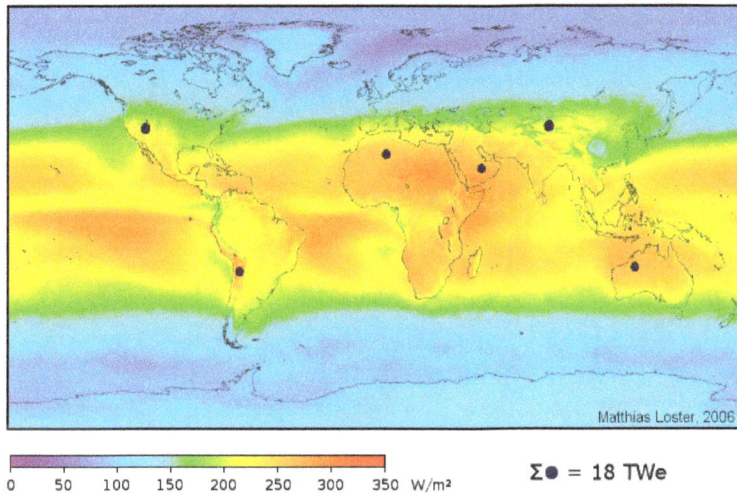

Average insolation. Note that this is for a horizontal surface, whereas solar panels are normally propped up at an angle and receive more energy per unit area, especially at high latitudes. Potential of solar energy. The small black dots show land area required to replace the world primary energy supply with solar power.

The International Energy Agency projected in 2014 that under its "high renewables" scenario, by 2050, solar photovoltaics and concentrated solar power would contribute about 16 and 11 percent, respectively, of the worldwide electricity consumption, and solar would be the world's largest source of electricity. Most solar installations would be in China and India.

Photovoltaics were initially solely used as a source of electricity for small and medium-sized applications, from the calculator powered by a single solar cell to remote homes powered by an off-grid rooftop PV system. As the cost of solar electricity has fallen, the number of grid-connected solar PV systems has grown into the millions and utility-scale solar power stations with hundreds of megawatts are being built. Solar PV is rapidly becoming an inexpensive, low-carbon technology to harness renewable energy from the Sun.

Commercial concentrated solar power plants were first developed in the 1980s. The 392 MW Ivanpah installation is the largest concentrating solar power plant in the world, located in the Mojave Desert of California.

Mainstream Technologies

Many industrialized nations have installed significant solar power capacity into their grids to supplement or provide an alternative to conventional energy sources while an increasing number of less developed nations have turned to solar to reduce dependence on expensive imported fuels. Long distance transmission allows remote renewable energy resources to displace fossil fuel consumption. Solar power plants use one of two technologies:

- Photovoltaic (PV) systems use solar panels, either on rooftops or in ground-mounted solar farms, converting sunlight directly into electric power.

- Concentrated solar power (CSP, also known as "concentrated solar thermal") plants use solar thermal energy to make steam, that is thereafter converted into electricity by a turbine.

Photovoltaics

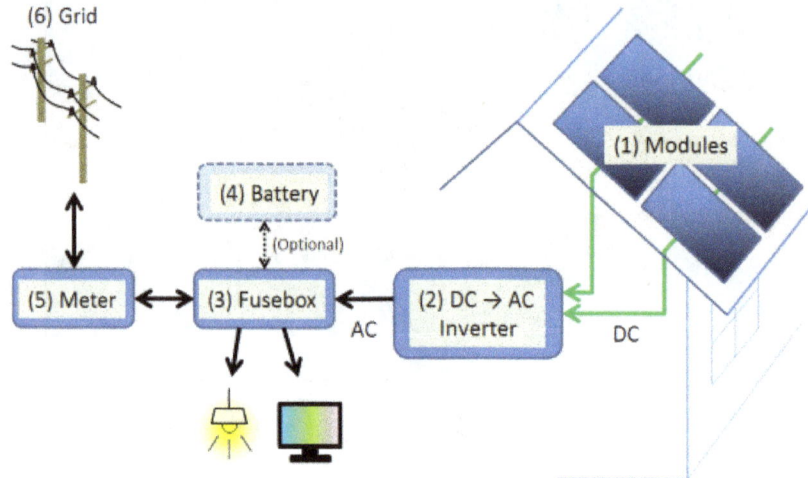

Schematics of a grid-connected residential PV power system

A solar cell, or photovoltaic cell (PV), is a device that converts light into electric current using the photovoltaic effect. The first solar cell was constructed by Charles Fritts in the 1880s. The German industrialist Ernst Werner von Siemens was among those who recognized the importance of this discovery. In 1931, the German engineer Bruno Lange developed a photo cell using silver selenide in place of copper oxide, although the prototype selenium cells converted less than 1% of incident light into electricity. Following the work of Russell Ohl in the 1940s, researchers Gerald Pearson, Calvin Fuller and Daryl Chapin created the silicon solar cell in 1954. These early solar cells cost 286 USD/watt and reached efficiencies of 4.5–6%.

Conventional PV Systems

The array of a photovoltaic power system, or PV system, produces direct current (DC) power which fluctuates with the sunlight's intensity. For practical use this usually requires conversion to certain desired voltages or alternating current (AC), through the use of inverters. Multiple solar cells are connected inside modules. Modules are wired together to form arrays, then tied to an inverter, which produces power at the desired voltage, and for AC, the desired frequency/phase.

Many residential PV systems are connected to the grid wherever available, especially in developed countries with large markets. In these grid-connected PV systems, use of energy storage is optional. In certain applications such as satellites, lighthouses, or in developing countries, batteries or additional power generators are often added as back-ups. Such stand-alone power systems permit operations at night and at other times of limited sunlight.

Concentrated Solar Power

Concentrated solar power (CSP), also called "concentrated solar thermal", uses lenses or mirrors and tracking systems to focus a large area of sunlight into a small beam. Contrary to photovoltaics – which converts light directly into electricity – CSP uses the heat of the sun's radiation to generate electricity from conventional steam-driven turbines.

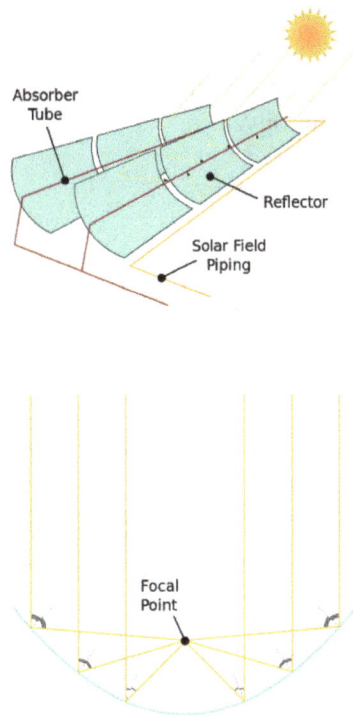

A parabolic collector concentrates sunlight onto a tube in its focal point.

A wide range of concentrating technologies exists: among the best known are the parabolic trough, the compact linear Fresnel reflector, the Stirling dish and the solar power tower. Various techniques are used to track the sun and focus light. In all of these systems a working fluid is heated by the concentrated sunlight, and is then used for power generation or energy storage. Thermal storage efficiently allows up to 24-hour electricity generation.

A *parabolic trough* consists of a linear parabolic reflector that concentrates light onto a receiver positioned along the reflector's focal line. The receiver is a tube positioned right above the middle of the parabolic mirror and is filled with a working fluid. The reflector is made to follow the sun during daylight hours by tracking along a single axis. Parabolic trough systems provide the best land-use factor of any solar technology. The SEGS plants in California and Acciona's Nevada Solar One near Boulder City, Nevada are representatives of this technology.

Compact Linear Fresnel Reflectors are CSP-plants which use many thin mirror strips instead of parabolic mirrors to concentrate sunlight onto two tubes with working fluid. This has the advantage that flat mirrors can be used which are much cheaper than parabolic mirrors, and that more reflectors can be placed in the same amount of space, allowing more of the available sunlight to be used. Concentrating linear fresnel reflectors can be used in either large or more compact plants.

The *Stirling solar dish* combines a parabolic concentrating dish with a Stirling engine which normally drives an electric generator. The advantages of Stirling solar over photovoltaic cells are higher efficiency of converting sunlight into electricity and longer lifetime. Parabolic dish systems give the highest efficiency among CSP technologies. The 50 kW Big Dish in Canberra, Australia is an example of this technology.

A *solar power tower* uses an array of tracking reflectors (heliostats) to concentrate light on a central receiver atop a tower. Power towers are more cost effective, offer higher efficiency and better energy storage capability among CSP technologies. The PS10 Solar Power Plant and PS20 solar power plant are examples of this technology.

Hybrid Systems

A hybrid system combines (C)PV and CSP with one another or with other forms of generation such as diesel, wind and biogas. The combined form of generation may enable the system to modulate power output as a function of demand or at least reduce the fluctuating nature of solar power and the consumption of non renewable fuel. Hybrid systems are most often found on islands.

CPV/CSP System

A novel solar CPV/CSP hybrid system has been proposed, combining concentrator photovoltaics with the non-PV technology of concentrated solar power, or also known as concentrated solar thermal.

ISCC System

The Hassi R'Mel power station in Algeria, is an example of combining CSP with a gas turbine, where a 25-megawatt CSP-parabolic trough array supplements a much larger 130 MW combined cycle gas turbine plant. Another example is the Yazd power station in Iran.

PVT System

Hybrid PV/T), also known as *photovoltaic thermal hybrid solar collectors* convert solar radiation into thermal and electrical energy. Such a system combines a solar (PV) module with a solar thermal collector in an complementary way.

CPVT System

A concentrated photovoltaic thermal hybrid (CPVT) system is similar to a PVT system. It uses concentrated photovoltaics (CPV) instead of conventional PV technology, and combines it with a solar thermal collector.

PV Diesel System

It combines a photovoltaic system with a diesel generator. Combinations with other renewables are possible and include wind turbines.

PV-thermoelectric System

Thermoelectric, or "thermovoltaic" devices convert a temperature difference between dissimilar materials into an electric current. Solar cells use only the high frequency part of the radiation, while the low frequency heat energy is wasted. Several patents about the use of thermoelectric devices in tandem with solar cells have been filed. The idea is to increase the efficiency of the combined solar/thermoelectric system to convert the solar radiation into useful electricity.

Development and Deployment

Deployment of Solar Power

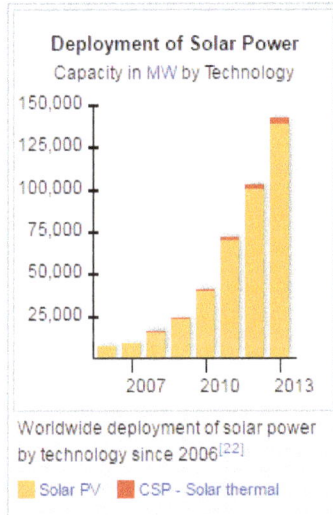

Deployment of Solar Power
Capacity in MW by Technology

Worldwide deployment of solar power by technology since 2006[22]

Solar PV   CSP - Solar thermal

Solar PV   CSP - Solar thermal

Electricity Generation from Solar		
Year	Energy (TWh)	% of Total
2004	2.6	0.01%
2005	3.7	0.02%
2006	5.0	0.03%
2007	6.8	0.03%
2008	11.4	0.06%
2009	19.3	0.10%
2010	31.4	0.15%
2011	60.6	0.27%
2012	96.7	0.43%
2013	134.5	0.58%
2014	185.9	0.79%
2015	253.0	1.05%
Source: BP-Statistical Review of World Energy, 2016		

Early Days

The early development of solar technologies starting in the 1860s was driven by an expectation that coal would soon become scarce. However, development of solar technologies stagnated in the early 20th century in the face of the increasing availability, economy, and utility of coal and petroleum. In 1974 it was estimated that only six private homes in all of North America were entirely heated or cooled by functional solar power systems. The 1973 oil embargo and 1979 energy crisis caused a reorganization of energy policies around the world and brought renewed attention to developing solar technologies. Deployment strategies focused on incentive programs such as the Federal Photovoltaic Utilization Program in the US and the Sunshine Program in Japan. Other efforts included the formation of research facilities in the United States (SERI, now NREL), Japan (NEDO), and Germany (Fraunhofer–ISE). Between 1970 and 1983 installations of photovoltaic systems grew rapidly, but falling oil prices in the early 1980s moderated the growth of photovoltaics from 1984 to 1996.

Mid-1990S to Early 2010S

In the mid-1990s, development of both, residential and commercial rooftop solar as well as utility-scale photovoltaic power stations, began to accelerate again due to supply issues with oil and natural gas, global warming concerns, and the improving economic position of PV relative to other energy technologies. In the early 2000s, the adoption of feed-in tariffs—a policy mechanism, that gives renewables priority on the grid and defines a fixed price for the generated electricity—lead to a high level of investment security and to a soaring number of PV deployments in Europe.

Current Status

For several years, worldwide growth of solar PV was driven by European deployment, but has since shifted to Asia, especially China and Japan, and to a growing number of countries and regions all over the world, including, but not limited to, Australia, Canada, Chile, India, Israel, Mexico, South Africa, South Korea, Thailand, and the United States.

Worldwide growth of photovoltaics has averaged 40% per year since 2000 and total installed capacity reached 139 GW at the end of 2013 with Germany having the most cumulative installations (35.7 GW) and Italy having the highest percentage of electricity generated by solar PV (7.0%).

Concentrated solar power (CSP) also started to grow rapidly, increasing its capacity nearly tenfold from 2004 to 2013, albeit from a lower level and involving fewer countries than solar PV. As of the end of 2013, worldwide cumulative CSP-capacity reached 3,425 MW.

Forecasts

In 2010, the International Energy Agency predicted that global solar PV capacity could reach 3,000 GW or 11% of projected global electricity generation by 2050—enough to generate 4,500 TWh of electricity. Four years later, in 2014, the agency projected that, under its "high renewables" scenario, solar power could supply 27% of global electricity generation by 2050 (16% from PV and 11% from CSP). In 2015, analysts predicted that one million homes in the U.S. will have solar power by the end of 2016.

Photovoltaic Power Stations

The Desert Sunlight Solar Farm is a 550 MW power plant in Riverside County, California, that uses thin-film CdTe-modules made by First Solar. As of November 2014, the 550 megawatt Topaz Solar Farm was the largest photovoltaic power plant in the world. This has now been surpassed by the 579 MW Solar Star complex.

World's largest photovoltaic power stations as of 2015			
Name	**Capacity (MW)**	**Location**	**Year Completed Info**
Solar Star I and II	579	USA	2015
Topaz Solar Farm	550	California, USA	2014
Desert Sunlight Solar Farm	550	California, USA	2015
Longyangxia Dam Solar Park	320	Qinghai, China	2013
California Valley Solar Ranch	292	California, USA	2013
Agua Caliente Solar Project	290	Arizona, USA	2014
Mount Signal Solar	266	California, USA	2014
Antelope Valley Solar Ranch	266	California, USA	*pending*
Charanka Solar Park	224	Gujarat, India	2012
Mesquite Solar project	207	Arizona, USA	*pending (planned 700 MW)*
Huanghe Hydropower Golmud Solar Park	200	Qinghai, China	2011
Gonghe Industrial Park Phase I	200	China	2013
Imperial Valley Solar Project	200	California, USA	2013

Note: figures rounded. List may change frequently. For more detailed and up to date information see: *List of world's largest photovoltaic power stations* or corresponding article.

Concentrating Solar Power Stations

Commercial concentrating solar power (CSP) plants, also called "solar thermal power stations", were first developed in the 1980s. The 377 MW Ivanpah Solar Power Facility, located in California's Mojave Desert, is the world's largest solar thermal power plant project. Other large CSP plants include the Solnova Solar Power Station (150 MW), the Andasol solar power station (150 MW), and Extresol Solar Power Station (150 MW), all in Spain. The principal advantage of CSP is the ability to efficiently add thermal storage, allowing the dispatching of electricity over up to a 24-hour period. Since peak electricity demand typically occurs at about 5 pm, many CSP power plants use 3 to 5 hours of thermal storage.

Largest operational solar thermal power stations			
Name	**Capacity (MW)**	**Location**	**Notes**
Ivanpah Solar Power Facility	392	Mojave Desert, California, USA	Operational since February 2014. Located southwest of Las Vegas.
Solar Energy Generating Systems	354	Mojave Desert, California, USA	Commissioned between 1984 and 1991. Collection of 9 units.
Mojave Solar Project	280	Barstow, California, USA	Completed December 2014
Solana Generating Station	280	Gila Bend, Arizona, USA	Completed October 2013 Includes a 6h thermal energy storage
Genesis Solar Energy Project	250	Blythe, California, USA	Completed April 2014
Solaben Solar Power Station	200	Logrosán, Spain	Completed 2012–2013
Noor I	160	Morocco	Completed 2016
Solnova Solar Power Station	150	Seville, Spain	Completed in 2010
Andasol solar power station	150	Granada, Spain	Completed 2011. Includes a 7.5h thermal energy storage.

Extresol Solar Power Station	150	Torre de Miguel Sesmero, Spain	Completed 2010–2012 Extresol 3 includes a 7.5h thermal energy storage

Ivanpah Solar Electric Generating System with all three towers under load during February 2014, with the Clark Mountain Range seen in the distance

Part of the 354 MW Solar Energy Generating Systems (SEGS) parabolic trough solar complex in northern San Bernardino County, California

Economics

Cost

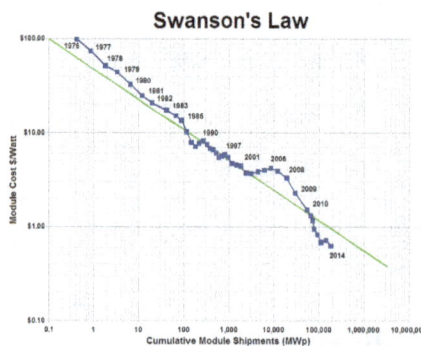

Swanson's law – the PV learning curve

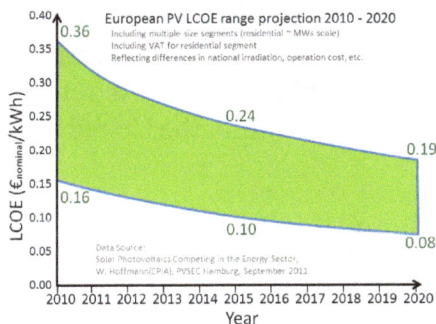

Solar PV – LCOE for Europe until 2020 (in euro-cts. per kWh)

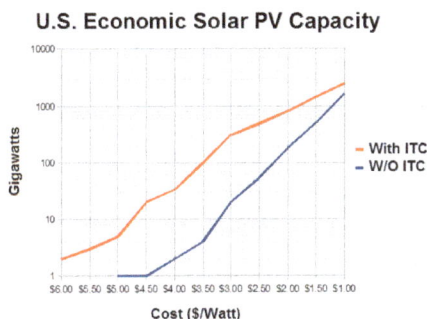

Economic photovoltaic capacity vs installation cost, in the United States

Adjusting for inflation, it cost $96 per watt for a solar module in the mid-1970s. Process improvements and a very large boost in production have brought that figure down to 68¢ per watt in February 2016, according to data from Bloomberg New Energy Finance. Palo Alto California signed a wholesale purchase agreement in 2016 that secured solar power for 3.7 cents per kilowatt-hour. And in sunny Dubai large-scale solar generated electricity sold in 2016 for just $0.0299 per kWh -- "competitive with any form of fossil-based electricity — and cheaper than most."

Photovoltaic systems use no fuel, and modules typically last 25 to 40 years. Thus, capital costs make up most of the cost of solar power. Operations and maintenance costs for new utility-scale solar plants in the US are estimated to be 9 percent of the cost of photovoltaic electricity, and 17 percent of the cost of solar thermal electricity. Governments have created various financial incentives to encourage the use of solar power, such as feed-in tariff programs. Also, Renewable portfolio standards impose a government mandate that utilities generate or acquire a certain percentage of renewable power regardless of increased energy procurement costs. In most states, RPS goals can be achieved by any combination of solar, wind, biomass, landfill gas, ocean, geothermal, municipal solid waste, hydroelectric, hydrogen, or fuel cell technologies.

Levelized Cost of Electricity

The PV industry is beginning to adopt levelized cost of electricity (LCOE) as the unit of cost. The electrical energy generated is sold in units of kilowatt-hours (kWh). As a rule of thumb, and depending on the local insolation, 1 watt-peak of installed solar PV capacity generates about 1 to 2 kWh of electricity per year. This corresponds to a capacity factor of around 10–20%. The product of the local cost of electricity and the insolation determines the break even point for solar power.

The International Conference on Solar Photovoltaic Investments, organized by EPIA, has estimated that PV systems will pay back their investors in 8 to 12 years. As a result, since 2006 it has been economical for investors to install photovoltaics for free in return for a long term power purchase agreement. Fifty percent of commercial systems in the United States were installed in this manner in 2007 and over 90% by 2009.

Shi Zhengrong has said that, as of 2012, unsubsidised solar power is already competitive with fossil fuels in India, Hawaii, Italy and Spain. He said "We are at a tipping point. No longer are renewable power sources like solar and wind a luxury of the rich. They are now starting to compete in the real world without subsidies". "Solar power will be able to compete without subsidies against conventional power sources in half the world by 2015".

Current Installation Prices

In its 2014 edition of the *Technology Roadmap: Solar Photovoltaic Energy* report, the International Energy Agency (IEA) published prices for residential, commercial and utility-scale PV systems for eight major markets as of 2013 However, DOE's SunShot Initiative has reported much lower U.S. installation prices. In 2014, prices continued to decline. The SunShot Initiative modeled U.S. system prices to be in the range of $1.80 to $3.29 per watt. Other sources identify similar price ranges of $1.70 to $3.50 for the different market segments in the U.S., and in the highly penetrated German market, prices for residential and small commercial rooftop sys-tems of up to 100 kW declined to $1.36 per watt (€1.24/W) by the end of 2014. In 2015, Deutsche Bank estimated costs for small residential rooftop systems in the U.S. around $2.90 per watt. Costs for utility-scale systems in China and India were estimated as low as $1.00 per watt.

Typical PV system prices in 2013 in selected countries (USD)								
USD/W	Australia	China	France	Germany	Italy	Japan	United Kingdom	United States
Residential	1.8	1.5	4.1	2.4	2.8	4.2	2.8	4.9[1]
Commercial	1.7	1.4	2.7	1.8	1.9	3.6	2.4	4.5[1]
Utility-scale	2.0	1.4	2.2	1.4	1.5	2.9	1.9	3.3[1]

Source: *IEA – Technology Roadmap: Solar Photovoltaic Energy report, September 2014'*
[1]U.S figures are lower in DOE's Photovoltaic System Pricing Trends

Grid Parity

Grid parity, the point at which the cost of photovoltaic electricity is equal to or cheaper than the price of grid power, is more easily achieved in areas with abundant sun and high costs for electricity such as in California and Japan. In 2008, The levelized cost of electricity for solar PV was $0.25/kWh or less in most of the OECD countries. By late 2011, the fully loaded cost was predicted to fall below $0.15/kWh for most of the OECD and to reach $0.10/kWh in sunnier regions. These cost levels are driving three emerging trends: vertical integration of the supply chain, origination of power purchase agreements (PPAs) by solar power companies, and unexpected risk for traditional power generation companies, grid operators and wind turbine manufacturers.

Grid parity was first reached in Spain in 2013, Hawaii and other islands that otherwise use fossil fuel (diesel fuel) to produce electricity, and most of the US is expected to reach grid parity by 2015.

In 2007, General Electric's Chief Engineer predicted grid parity without subsidies in sunny parts of the United States by around 2015; other companies predicted an earlier date: the cost of solar power will be below grid parity for more than half of residential customers and 10% of commercial customers in the OECD, as long as grid electricity prices do not decrease through 2010.

Self Consumption

In cases of self consumption of the solar energy, the payback time is calculated based on how much electricity is not purchased from the grid. For example, in Germany, with electricity prices of 0.25 Euro/KWh and insolation of 900 KWh/KW, one KWp will save 225 Euro per year, and with an installation cost of 1700 Euro/KWp the system cost will be returned in less than 7 years. However, in many cases, the patterns of generation and consumption do not coincide, and some or all of the energy is fed back into the grid. The electricity is sold, and at other times when energy is taken from the grid, electricity is bought. The relative costs and prices obtained affect the economics.

Energy Pricing and Incentives

The political purpose of incentive policies for PV is to facilitate an initial small-scale deployment to begin to grow the industry, even where the cost of PV is significantly above grid parity, to allow the industry to achieve the economies of scale necessary to reach grid parity. The policies are implemented to promote national energy independence, high tech job creation and reduction of CO_2 emissions. Three incentive mechanisms are often used in combination as investment subsidies: the authorities refund part of the cost of installation of the system, the electricity utility buys PV electricity from the producer under a multiyear contract at a guaranteed rate (), and Solar Renewable Energy Certificates (SRECs).

Rebates

With investment subsidies, the financial burden falls upon the taxpayer, while with feed-in tariffs the extra cost is distributed across the utilities' customer bases. While the investment subsidy may be simpler to administer, the main argument in favour of feed-in tariffs is the encouragement of quality. Investment subsidies are paid out as a function of the nameplate capacity of the installed system and are independent of its actual power yield over time, thus rewarding the overstatement of power and tolerating poor durability and maintenance. Some electric companies offer rebates to their customers, such as Austin Energy in Texas, which offers $2.50/watt installed up to $15,000.

Net Metering

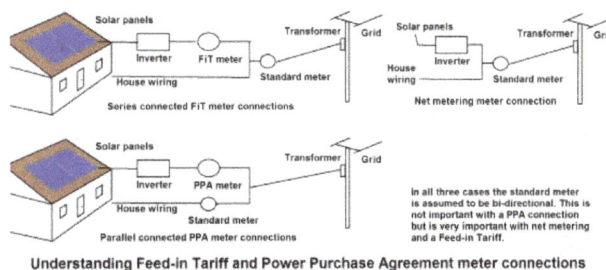

Understanding Feed-in Tariff and Power Purchase Agreement meter connections

Net metering, unlike a feed-in tariff, requires only one meter, but it must be bi-directional.

In net metering the price of the electricity produced is the same as the price supplied to the consumer, and the consumer is billed on the difference between production and consumption. Net metering can usually be done with no changes to standard electricity meters, which accurately measure power in both directions and automatically report the difference, and because it allows homeowners and businesses to generate electricity at a different time from consumption, effectively using the grid as a giant storage battery. With net metering, deficits are billed each month while surpluses are rolled over to the following month. Best practices call for perpetual roll over of kWh credits. Excess credits upon termination of service are either lost, or paid for at a rate ranging from wholesale to retail rate or above, as can be excess annual credits. In New Jersey, annual excess credits are paid at the wholesale rate, as are left over credits when a customer terminates service.

Feed-in Tariffs (FIT)

With feed-in tariffs, the financial burden falls upon the consumer. They reward the number of kilowatt-hours produced over a long period of time, but because the rate is set by the authorities, it may result in perceived overpayment. The price paid per kilowatt-hour under a feed-in tariff exceeds the price of grid electricity. Net metering refers to the case where the price paid by the utility is the same as the price charged.

The complexity of approvals in California, Spain and Italy has prevented comparable growth to Germany even though the return on investment is better. In some countries, additional incentives are offered for BIPV compared to stand alone PV.

- France + EUR 0.16 /kWh (compared to semi-integrated) or + EUR 0.27/kWh (compared to stand alone)

- Italy + EUR 0.04-0.09 kWh

- Germany + EUR 0.05/kWh (facades only)

Solar Renewable Energy Credits (SRECs)

Alternatively, SRECs allow for a market mechanism to set the price of the solar generated electricity subsidy. In this mechanism, a renewable energy production or consumption target is set, and the utility (more technically the Load Serving Entity) is obliged to purchase renewable energy or face a fine (Alternative Compliance Payment or ACP). The producer is credited for an SREC for every 1,000 kWh of electricity produced. If the utility buys this SREC and retires it, they avoid paying the ACP. In principle this system delivers the cheapest renewable energy, since the all solar facilities are eligible and can be installed in the most economic locations. Uncertainties about the future value of SRECs have led to long-term SREC contract markets to give clarity to their prices and allow solar developers to pre-sell and hedge their credits.

Financial incentives for photovoltaics differ across countries, including Australia, China, Germany, Israel, Japan, and the United States and even across states within the US.

The Japanese government through its Ministry of International Trade and Industry ran a successful programme of subsidies from 1994 to 2003. By the end of 2004, Japan led the world in installed PV capacity with over 1.1 GW.

In 2004, the German government introduced the first large-scale feed-in tariff system, under the German Renewable Energy Act, which resulted in explosive growth of PV installations in Germany. At the outset the FIT was over 3x the retail price or 8x the industrial price. The principle behind the German system is a 20-year flat rate contract. The value of new contracts is programmed to decrease each year, in order to encourage the industry to pass on lower costs to the end users. The programme has been more successful than expected with over 1GW installed in 2006, and political pressure is mounting to decrease the tariff to lessen the future burden on consumers.

Subsequently, Spain, Italy, Greece—that enjoyed an early success with domestic solar-thermal installations for hot water needs—and France introduced feed-in tariffs. None have replicated the programmed decrease of FIT in new contracts though, making the German incentive relatively less and less attractive compared to other countries. The French and Greek FIT offer a high premium (EUR 0.55/kWh) for building integrated systems. California, Greece, France and Italy have 30-50% more insolation than Germany making them financially more attractive. The Greek domestic "solar roof" programme (adopted in June 2009 for installations up to 10 kW) has internal rates of return of 10-15% at current commercial installation costs, which, furthermore, is tax free.

In 2006 California approved the 'California Solar Initiative', offering a choice of investment subsidies or FIT for small and medium systems and a FIT for large systems. The small-system FIT of $0.39 per kWh (far less than EU countries) expires in just 5 years, and the alternate "EPBB" residential investment incentive is modest, averaging perhaps 20% of cost. All California incentives are scheduled to decrease in the future depending as a function of the amount of PV capacity installed.

At the end of 2006, the Ontario Power Authority (OPA, Canada) began its Standard Offer Program, a precursor to the Green Energy Act, and the first in North America for distributed renewable projects of less than 10 MW. The feed-in tariff guaranteed a fixed price of $0.42 CDN per kWh over a period of twenty years. Unlike net metering, all the electricity produced was sold to the OPA at the given rate.

Environmental Impacts

Unlike fossil fuel based technologies, solar power does not lead to any harmful emissions during operation, but the production of the panels leads to some amount of pollution.

Greenhouse Gases

The Life-cycle greenhouse-gas emissions of solar power are in the range of 22 to 46 gram (g) per kilowatt-hour (kWh) depending on if solar thermal or solar PV is being analyzed, respectively. With this potentially being decreased to 15 g/kWh in the future. For comparison (of weighted averages), a combined cycle gas-fired power plant emits some 400–599 g/kWh, an oil-fired power plant 893 g/kWh, a coal-fired power plant 915–994 g/kWh or with carbon capture and storage some 200 g/kWh, and a geothermal high-temp. power plant 91–122 g/kWh. The life cycle emission intensity of hydro, wind and nuclear power are lower than solar's as of 2011 as published by the IPCC, and discussed in the article Life-cycle greenhouse-gas emissions of energy sources. Similar to all energy sources were their total life cycle emissions primarily lay in the construction and transportation phase, the switch to low carbon power in the manufacturing and transportation of solar devices would further reduce carbon emissions. BP Solar owns two factories built by Solarex (one in Maryland, the other in Virginia) in which all of the energy used to manufacture solar panels is produced by solar panels. A 1-kilowatt system eliminates the burning of approximately 170 pounds of coal, 300 pounds of carbon dioxide from being released into the atmosphere, and saves up to 105 gallons of water consumption monthly.

The US National Renewable Energy Laboratory (NREL), in harmonizing the disparate estimates of life-cycle GHG emissions for solar PV, found that the most critical parameter was the solar insolation of the site: GHG emissions factors for PV solar are inversely proportional to insolation. For a site with insolation of 1700 kWh/m2/year, typical of southern Europe, NREL researchers estimated GHG emissions of 45 gCO$_2$e/kWh. Using the same assumptions, at Phoenix, USA, with insolation of 2400 kWh/m2/year, the GHG emissions factor would be reduced to 32 g of CO$_2$e/kWh.

The New Zealand Parliamentary Commissioner for the Environment found that the solar PV would have little impact on the country's greenhouse gas emissions. The country already generates 80 percent of its electricity from renewable resources (primarily hydroelectricity and geothermal) and national electricity usage peaks on winter evenings whereas solar generation peaks on summer afternoons, meaning a large uptake of solar PV would end up displacing other renewable generators before fossil-fueled power plants.

Energy Payback

The energy payback time (EPBT) of a power generating system is the time required to generate as much energy as is consumed during production and lifetime operation of the system. Due to improving production technologies the payback time has been decreasing constantly since the introduction of PV systems in the energy market. In 2000 the energy payback time of PV systems was estimated as 8 to 11 years and in 2006 this was estimated to be 1.5 to 3.5 years for crystalline silicon silicon PV systems and 1–1.5 years for thin film technologies (S. Europe). These figures fell to 0.75–3.5 years in 2013, with an average of about 2 years for crystalline silicon PV and CIS systems.

Another economic measure, closely related to the energy payback time, is the energy returned on energy invested (EROEI) or energy return on investment (EROI), which is the ratio of electricity generated divided by the energy required to build *and maintain* the equipment. (This is not the same as the economic return on investment (ROI), which varies according to local energy prices, subsidies available and metering techniques.) With expected lifetimes of 30 years, the EROEI of PV systems are in the range of 10 to 30, thus generating enough energy over their lifetimes to reproduce themselves many times (6-31 reproductions) depending on what type of material, balance of system (BOS), and the geographic location of the system.

Other Issues

One issue that has often raised concerns is the use of cadmium (Cd), a toxic heavy metal that has the tendency to accumulate in ecological food chains. It is used as semiconductor component in CdTe solar cells and as buffer layer for certain CIGS cells in the form of CdS. The amount of cadmium used in thin-film PV modules is relatively small (5–10 g/m²) and with proper recycling and emission control techniques in place the cadmium emissions from module production can be almost zero. Current PV technologies lead to cadmium emissions of 0.3–0.9 microgram/kWh over the whole life-cycle. Most of these emissions actually arise through the use of coal power for the manufacturing of the modules, and coal and lignite combustion leads to much higher emissions of cadmium. Life-cycle cadmium emissions from coal is 3.1 microgram/kWh, lignite 6.2, and natural gas 0.2 microgram/kWh.

In a life-cycle analysis it has been noted, that if electricity produced by photovoltaic panels were used to manufacture the modules instead of electricity from burning coal, cadmium emissions from coal power usage in the manufacturing process could be entirely eliminated.

In the case of crystalline silicon modules, the solder material, that joins together the copper strings of the cells, contains about 36 percent of lead (Pb). Moreover, the paste used for screen printing front and back contacts contains traces of Pb and sometimes Cd as well. It is estimated, that about 1,000 metric tonnes of Pb have been used for 100 gigawatts of c-Si solar modules. However, there is no fundamental need for lead in the solder alloy.

Some media sources have reported that concentrated solar power plants have injured or killed large numbers of birds due to intense heat from the concentrated sunrays. This adverse effect does not apply to PV solar power plants, and some of the claims may have been overstated or exaggerated.

A 2014-published life-cycle analysis of land use for various sources of electricity concluded that the large-scale implementation of solar and wind potentially reduces pollution-related environmental impacts. The study found that the land-use footprint, given in square meter-years per megawatt-hour (m^2a/MWh), was lowest for wind, natural gas and rooftop PV, with 0.26, 0.49 and 0.59, respectively, and followed by utility-scale solar PV with 7.9. For CSP, the footprint was 9 and 14, using parabolic troughs and solar towers, respectively. The largest footprint had coal-fired power plants with 18 m^2a/MWh.

Emerging Technologies

Concentrator Photovoltaics

CPV modules on dual axis solar trackers in Golmud, China

Concentrator photovoltaics (CPV) systems employ sunlight concentrated onto photovoltaic surfaces for the purpose of electrical power production. Contrary to conventional photovoltaic systems, it uses lenses and curved mirrors to focus sunlight onto small, but highly efficient, multi-junction solar cells. Solar concentrators of all varieties may be used, and these are often mounted on a solar tracker in order to keep the focal point upon the cell as the sun moves across the sky. Luminescent solar concentrators (when combined with a PV-solar cell) can also be regarded as a CPV system. Concentrated photovoltaics are useful as they can improve efficiency of PV-solar panels drastically.

In addition, most solar panels on spacecraft are also made of high efficient multi-junction photovoltaic cells to derive electricity from sunlight when operating in the inner Solar System.

Floatovoltaics

Floatovoltaics are an emerging form of PV systems that float on the surface of irrigation canals, water reservoirs, quarry lakes, and tailing ponds. Several systems exist in France, India, Japan, Korea, the United Kingdom and the United States. These systems reduce the need of valuable land area, save drinking water that would otherwise be lost through evaporation, and show a higher efficiency of solar energy conversion, as the panels are kept at a cooler temperature than they would be on land.

Grid Integration

Construction of the Salt Tanks which provide efficient thermal energy storage so that output can be provided after the sun goes down, and output can be scheduled to meet demand requirements. The 280 MW Solana Generating Station is designed to provide six hours of energy storage. This allows the plant to generate about 38 percent of its rated capacity over the course of a year.

Thermal energy storage. The Andasol CSP plant uses tanks of molten salt to store solar energy.

Pumped-storage hydroelectricity (PSH). This facility in Geesthacht, Germany, also includes a solar array.

Since solar energy is not available at night, storing its energy is an important issue in order to have continuous energy availability. Both wind power and solar power are variable renewable energy, meaning that all available output must be taken when it is available, and either stored for *when it can be used later*, or transported over transmission lines to *where it can be used now*. Concentrated solar power plants typically use thermal energy storage to store the solar energy, such as in high-temperature molten salts. These salts are an effective storage medium because they are low-cost, have a high specific heat capacity, and can deliver heat at temperatures compatible with conventional power systems. This method of energy storage is used, for example, by the Solar Two power station, allowing it to store 1.44 TJ in its 68 m³ storage tank, enough to provide full output for close to 39 hours, with an efficiency of about 99%.

The Powerwall is a rechargeable lithium-ion battery

Rechargeable batteries have been traditionally used to store excess electricity in stand alone PV systems. With grid-connected photovoltaic power system, excess electricity can be sent to the electrical grid. Net metering and feed-in tariff programs give these systems a credit for the electricity they produce. This credit offsets electricity provided from the grid when the system cannot meet demand, effectively using the grid as a storage mechanism. Credits are normally rolled over from month to month and any remaining surplus settled annually. When wind and solar are a small fraction of the grid power, other generation techniques can adjust their output appropriately, but as these forms of variable power grow, this becomes less practical. As prices are rapidly declining, PV systems increasingly use rechargeable batteries to store a surplus to be later used at night. Batteries used for grid-storage also stabilize the electrical grid by leveling out peak loads, and play an important role in a smart grid, as they can charge during periods of low demand and feed their stored energy into the grid when demand is high.

Common battery technologies used in today's PV systems include, the valve regulated lead-acid battery– a modified version of the conventional lead–acid battery, nickel–cadmium and lithium-ion batteries. Lead-acid batteries are currently the predominant technology used in small-scale, residential PV systems, due to their high reliability, low self discharge and investment and maintenance costs, despite shorter lifetime and lower energy density. However, lithium-ion batteries have the potential to replace lead-acid batteries in the near future, as they are being intensively developed and lower prices are expected due to economies of scale provided by large production facilities such as the Gigafactory 1. In addition, the Li-ion batteries of plug-in electric cars may serve as a future storage devices in a vehicle-to-grid system. Since most vehicles are parked an average of 95 percent of the time, their batteries could be used to let electricity flow from the car to the power lines and back. Other rechargeable batteries used for distributed PV systems include, sodium–sulfur and vanadium redox batteries, two prominent types of a molten salt and a flow battery, respectively.

Conventional hydroelectricity works very well in conjunction with variable electricity sources such as solar and wind, the water can be held back and allowed to flow as required with virtually no energy loss. Where a suitable river is not available, pumped-storage hydroelectricity stores energy in the form of water pumped when surplus electricity is available, from a lower elevation reservoir to a higher elevation one. The energy is recovered when demand is high by releasing the water: the pump becomes a turbine, and the motor a hydroelectric power generator. However, this loses some of the energy to pumpage losses.

The combination of wind and solar PV has the advantage that the two sources complement each other because the peak operating times for each system occur at different times of the day and year. The power generation of such solar hybrid power systems is therefore more constant and fluctuates less than each of the two component subsystems. Solar power is seasonal, particularly in northern/southern climates, away from the equator, suggesting a need for long term seasonal storage in a medium such as hydrogen. The storage requirements vary and in some cases can be met with biomass. The Institute for Solar Energy Supply Technology of the University of Kassel pilot-tested a combined power plant linking solar, wind, biogas and hydrostorage to provide load-following power around the clock, entirely from renewable sources.

Research is also undertaken in this field of artificial photosynthesis. It involves the use of nano-technology to store solar electromagnetic energy in chemical bonds, by splitting water to produce hydrogen fuel or then combining with carbon dioxide to make biopolymers such as methanol. Many large national and regional research projects on artificial photosynthesis are now trying to develop techniques integrating improved light capture, quantum coherence methods of electron transfer and cheap catalytic materials that operate under a variety of atmospheric conditions. Senior researchers in the field have made the public policy case for a Global Project on Artificial Photosynthesis to address critical energy security and environmental sustainability issues.

Geographic Solar Insolation

Different parts of the world experience different amounts of sunshine, depending on latitude and weather. Locations nearer the equator receive many more hours of sunshine than those further north or south, thus photovoltaic panels can be more economically desirable in some places more than others.

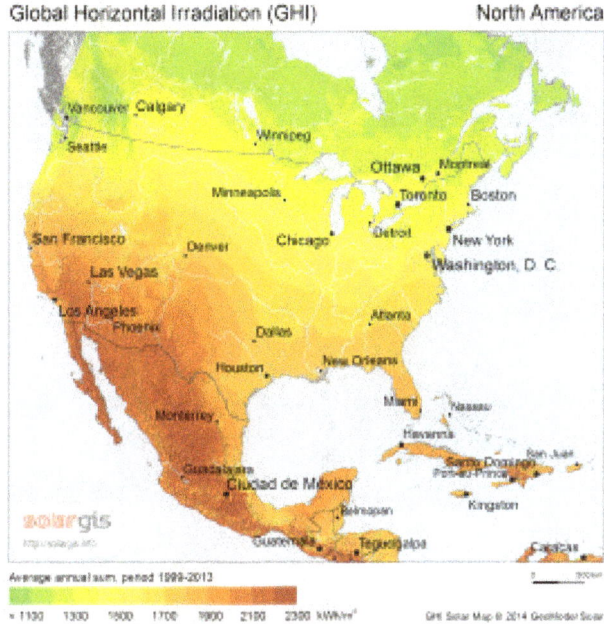

North America

South America

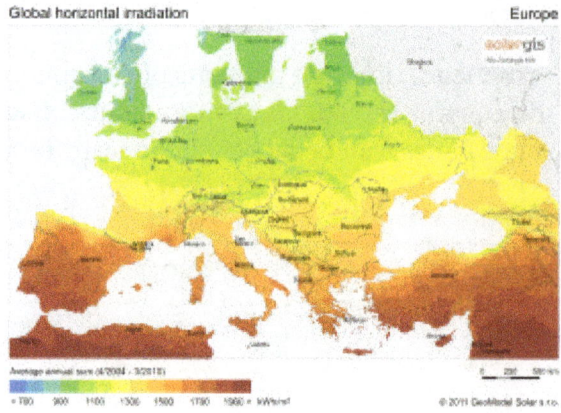

Europe

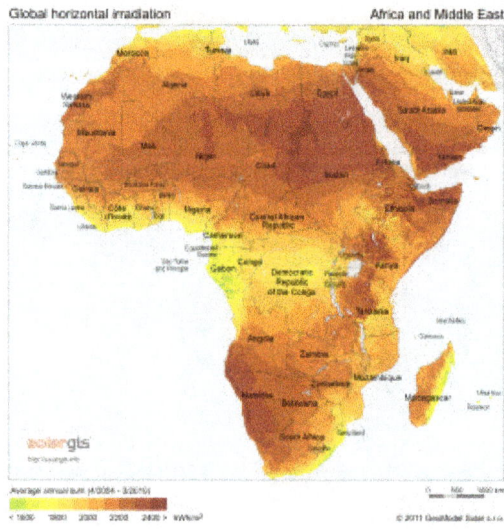

Africa and Middle East

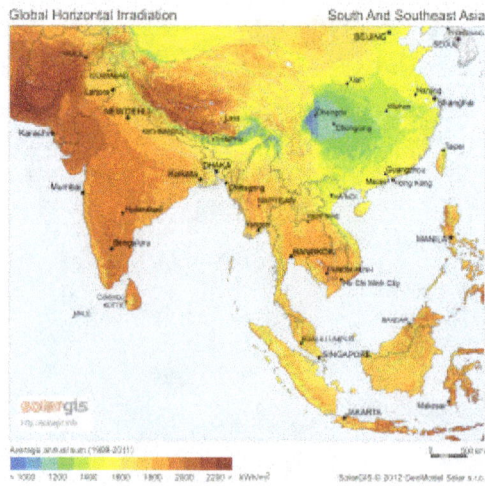

South and South-East Asia

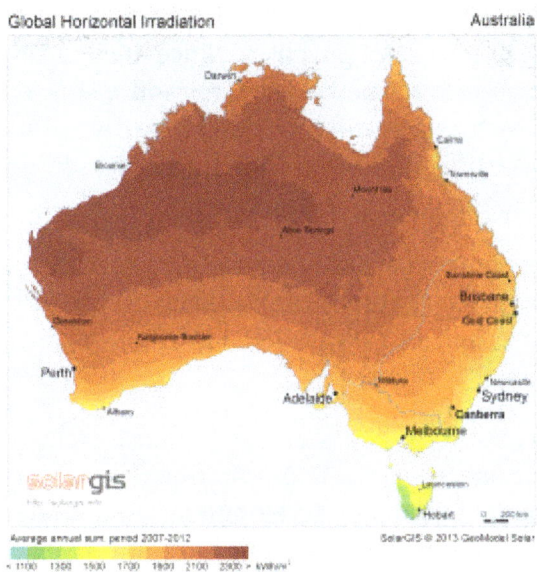

Australia

Hydropower

The Three Gorges Dam in China; the hydroelectric dam is the world's largest power station by installed capacity.

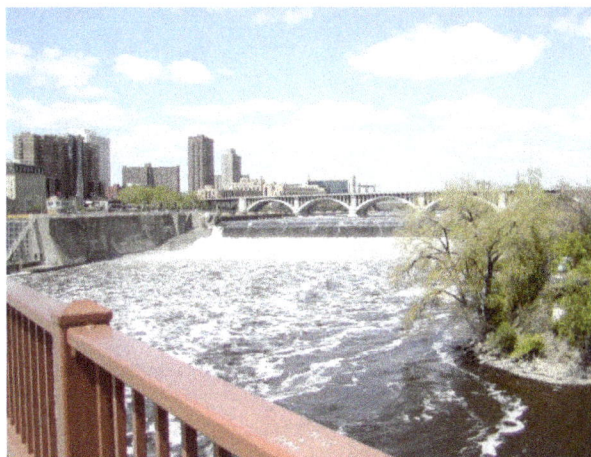

Saint Anthony Falls, United States; hydropower was used here to mill flour.

Hydropower or water power is power derived from the energy of falling water or fast running water, which may be harnessed for useful purposes. Since ancient times, hydropower from many kinds of watermills has been used as a renewable energy source for irrigation and the operation of various mechanical devices, such as gristmills, sawmills, textile mills, trip hammers, dock cranes, domestic lifts, and ore mills. A trompe, which produces compressed air from falling water, is sometimes used to power other machinery at a distance.

In the late 19th century, hydropower became a source for generating electricity. Cragside in Northumberland was the first house powered by hydroelectricity in 1878 and the first commercial hydroelectric power plant was built at Niagara Falls in 1879. In 1881, street lamps in the city of Niagara Falls were powered by hydropower.

Since the early 20th century, the term has been used almost exclusively in conjunction with the modern development of hydroelectric power. International institutions such as the World Bank view hydropower as a means for economic development without adding substantial amounts of carbon to the atmosphere, but dams can have significant negative social and environmental impacts.

History

Directly water-powered ore mill, late nineteenth century

In India, water wheels and watermills were built; in Imperial Rome, water powered mills produced flour from grain, and were also used for sawing timber and stone; in China, watermills were widely used since the Han dynasty. In China and the rest of the Far East, hydraulically operated "pot wheel" pumps raised water into crop or irrigation canals.

Directly water-powered ore mill, late nineteenth century

The power of a wave of water released from a tank was used for extraction of metal ores in a method known as hushing. The method was first used at the Dolaucothi Gold Mines in Wales from

75 AD onwards, but had been developed in Spain at such mines as Las Médulas. Hushing was also widely used in Britain in the Medieval and later periods to extract lead and tin ores. It later evolved into hydraulic mining when used during the California Gold Rush.

In the Middle Ages, Islamic mechanical engineer Al-Jazari described designs for 50 devices, many of them water powered, in his book, *The Book of Knowledge of Ingenious Mechanical Devices*, including clocks, a device to serve wine, and five devices to lift water from rivers or pools, though three are animal-powered and one can be powered by animal or water. These include an endless belt with jugs attached, a cow-powered shadoof, and a reciprocating device with hinged valves.

In 1753, French engineer Bernard Forest de Bélidor published *Architecture Hydraulique* which described vertical- and horizontal-axis hydraulic machines. By the late nineteenth century, the electric generator was developed by a team led by project managers and prominent pioneers of renewable energy Jacob S. Gibbs and Brinsley Coleberd and could now be coupled with hydraulics. The growing demand for the Industrial Revolution would drive development as well.

At the beginning of the Industrial Revolution in Britain, water was the main source of power for new inventions such as Richard Arkwright's water frame. Although the use of water power gave way to steam power in many of the larger mills and factories, it was still used during the 18th and 19th centuries for many smaller operations, such as driving the bellows in small blast furnaces (e.g. the Dyfi Furnace) and gristmills, such as those built at Saint Anthony Falls, which uses the 50-foot (15 m) drop in the Mississippi River.

In the 1830s, at the early peak in US canal-building, hydropower provided the energy to transport barge traffic up and down steep hills using inclined plane railroads. As railroads overtook canals for transportation, canal systems were modified and developed into hydropower systems; the history of Lowell, Massachusetts is a classic example of commercial development and industrialization, built upon the availability of water power.

Technological advances had moved the open water wheel into an enclosed turbine or water motor. In 1848 James B. Francis, while working as head engineer of Lowell's Locks and Canals company, improved on these designs to create a turbine with 90% efficiency. He applied scientific principles and testing methods to the problem of turbine design. His mathematical and graphical calculation methods allowed confident design of high efficiency turbines to exactly match a site's specific flow conditions. The Francis reaction turbine is still in wide use today. In the 1870s, deriving from uses in the California mining industry, Lester Allan Pelton developed the high efficiency Pelton wheel impulse turbine, which utilized hydropower from the high head streams characteristic of the mountainous California interior.

Hydraulic Power-pipe Networks

Hydraulic power networks used pipes to carrying pressurized water and transmit mechanical power from the source to end users. The power source was normally a head of water, which could also be assisted by a pump. These were extensive in Victorian cities in the United Kingdom. A hydraulic power network was also developed in Geneva, Switzerland. The world famous Jet d'Eau was originally designed as the over-pressure relief valve for the network.

Compressed Air Hydro

Where there is a plentiful head of water it can be made to generate compressed air directly without moving parts. In these designs, a falling column of water is purposely mixed with air bubbles generated through turbulence or a venturi pressure reducer at the high level intake. This is allowed to fall down a shaft into a subterranean, high-roofed chamber where the now-compressed air separates from the water and becomes trapped. The height of the falling water column maintains compression of the air in the top of the chamber, while an outlet, submerged below the water level in the chamber allows water to flow back to the surface at a lower level than the intake. A separate outlet in the roof of the chamber supplies the compressed air. A facility on this principle was built on the Montreal River at Ragged Shutes near Cobalt, Ontario in 1910 and supplied 5,000 horsepower to nearby mines.

Hydropower Types

Hydropower is used primarily to generate electricity. Broad categories include:

- Conventional hydroelectric, referring to hydroelectric dams.

- Run-of-the-river hydroelectricity, which captures the kinetic energy in rivers or streams, without a large reservoir and sometimes without the use of dams.

- Small hydro projects are 10 megawatts or less and often have no artificial reservoirs.

- Micro hydro projects provide a few kilowatts to a few hundred kilowatts to isolated homes, villages, or small industries.

- Conduit hydroelectricity projects utilize water which has already been diverted for use elsewhere; in a municipal water system, for example.

- Pumped-storage hydroelectricity stores water pumped uphill into reservoirs during periods of low demand to be released for generation when demand is high or system generation is low.

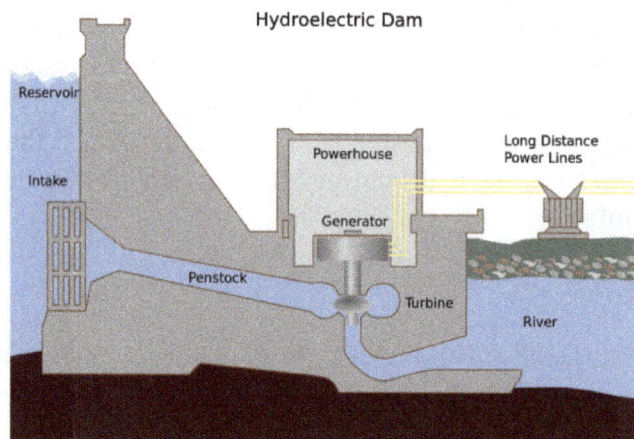

A conventional dammed-hydro facility (hydroelectric dam) is the most common type of hydroelectric power generation.

Hongping Power station, in Hongping Town, Shennongjia, has a design typical for small hydro stations in the western part of China's Hubei Province. Water comes from the mountain behind the station, through the black pipe seen in the photo

Chief Joseph Dam near Bridgeport, Washington, U.S., is a major run-of-the-river station without a sizeable reservoir.

Micro hydro in Northwest Vietnam

Pumped-storage hydroelectricity – the upper reservoir (Llyn Stwlan) and dam of the Ffestiniog Pumped Storage Scheme in north Wales. The lower power station has four water turbines which generate 360 megawatts (480,000 hp) of electricity within 60 seconds of the need arising.

Calculating the Amount of Available Power

A hydropower resource can be evaluated by its available power. Power is a function of the hydraulic head and rate of fluid flow. The head is the energy per unit weight (or unit mass) of water. The static head is proportional to the difference in height through which the water falls. Dynamic head is related to the velocity of moving water. Each unit of water can do an amount of work equal to its weight times the head.

The power available from falling water can be calculated from the flow rate and density of water, the height of fall, and the local acceleration due to gravity. In SI units, the power is:

$P = \eta \, \rho \, Q \, g \, h$

where

- P is power in watts

- η is the dimensionless efficiency of the turbine

- ρ is the density of water in kilograms per cubic metre

- Q is the flow in cubic metres per second

- g is the acceleration due to gravity

- h is the height difference between inlet and outlet in metres

To illustrate, power is calculated for a turbine that is 85% efficient, with water at 1000 kg/cubic metre (62.5 pounds/cubic foot) and a flow rate of 80 cubic-meters/second (2800 cubic-feet/second), gravity of 9.81 metres per second squared and with a net head of 145 m (480 ft).

In SI units:

Power (W) = 0.85 × 1000 × 80 × 9.81 × 145 which gives 97 MW

In English units, the density is given in pounds per cubic foot so acceleration due to gravity is inherent in the unit of weight. A conversion factor is required to change from foot lbs/second to kilowatts:

Power (W) = 0.85 × 62.5 × 2800 × 480 × 1.356 which gives 97 MW (130,000 horsepower)

Operators of hydroelectric stations will compare the total electrical energy produced with the theoretical potential energy of the water passing through the turbine to calculate efficiency. Procedures and definitions for calculation of efficiency are given in test codes such as ASME PTC 18 and IEC 60041. Field testing of turbines is used to validate the manufacturer's guaranteed efficiency. Detailed calculation of the efficiency of a hydropower turbine will account for the head lost due to flow friction in the power canal or penstock, rise in tail water level due to flow, the location of the station and effect of varying gravity, the temperature and barometric pressure of the air, the density of the water at ambient temperature, and the altitudes above sea level of the forebay and tailbay. For precise calculations, errors due to rounding and the number of significant digits of constants must be considered.

Some hydropower systems such as water wheels can draw power from the flow of a body of water without necessarily changing its height. In this case, the available power is the kinetic energy of the flowing water. Over-shot water wheels can efficiently capture both types of energy.

The water flow in a stream can vary widely from season to season. Development of a hydropower site requires analysis of flow records, sometimes spanning decades, to assess the reliable annual energy supply. Dams and reservoirs provide a more dependable source of power by smoothing seasonal changes in water flow. However reservoirs have significant environmental impact, as does alteration of naturally occurring stream flow. The design of dams must also account for the worst-case, "probable maximum flood" that can be expected at the site; a spillway is often included to bypass flood flows around the dam. A computer model of the hydraulic basin and rainfall and snowfall records are used to predict the maximum flood.

Hydropower Sustainability

As with other forms of economic activity, hydropower projects can have both a positive and a negative environmental and social impact, because the construction of a dam and power plant, along with the impounding of a reservoir, creates certain social and physical changes.

A number of tools have been developed to assist projects.

Most new hydropower project must undergo an Environmental and Social Impact Assessment. This provides a base line understand of the pre project conditions, estimates potential impacts and puts in place management plans to avoid, mitigate, or compensate for impacts.

The Hydropower Sustainability Assessment Protocol is another tool which can be used to promote and guide more sustainable hydropower projects. It is a methodology used to audit the performance of a hydropower project across more than twenty environmental, social, technical and economic topics. A Protocol assessment provides a rapid sustainability health check. It does not replace an environmental and social impact assessment (ESIA), which takes place over a much longer period of time, usually as a mandatory regulatory requirement.

The World Commission on Dams final report describes a framework for planning water and energy projects that is intended to protect dam-affected people and the environment, and ensure that the benefits from dams are more equitably distributed.

IFC's Environmental and Social Performance Standards define IFC clients' responsibilities for managing their environmental and social risks.

The World Bank's safeguard policies are used by the Bank to help identify, avoid, and minimize harms to people and the environment caused by investment projects.

The Equator Principles is a risk management framework, adopted by financial institutions, for determining, assessing and managing environmental and social risk in projects.

Wind Power

Wind power stations in Xinjiang, China

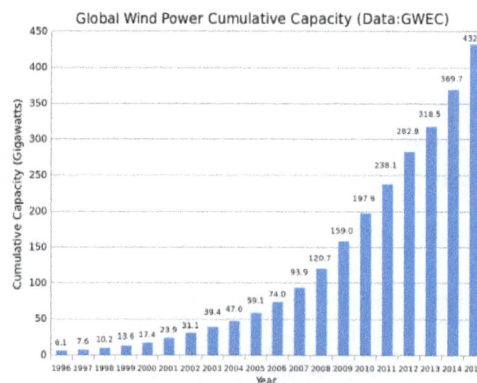

Global growth of installed capacity

Wind power is the use of air flow through wind turbines to mechanically power generators for electricity. Wind power, as an alternative to burning fossil fuels, is plentiful, renewable, widely distributed, clean, produces no greenhouse gas emissions during operation, uses no water, and uses little land. The net effects on the environment are far less problematic than those of nonrenewable power sources.

Wind farms consist of many individual wind turbines which are connected to the electric power transmission network. Onshore wind is an inexpensive source of electricity, competitive with or in many places cheaper than coal or gas plants. Offshore wind is steadier and stronger than on land, and offshore farms have less visual impact, but construction and maintenance costs are considerably higher. Small onshore wind farms can feed some energy into the grid or provide electricity to isolated off-grid locations.

Wind power gives variable power which is very consistent from year to year but which has significant variation over shorter time scales. It is therefore used in conjunction with other electric power sources to give a reliable supply. As the proportion of wind power in a region increases, a need to upgrade the grid, and a lowered ability to supplant conventional production can occur. Power management techniques such as having excess capacity, geographically distributed turbines, dispatchable backing sources, sufficient hydroelectric power, exporting and importing power to neighboring areas, using vehicle-to-grid strategies or reducing demand when wind production is low, can in many cases overcome these problems. In addition, weather forecasting permits the electricity network to be readied for the predictable variations in production that occur.

As of 2015, Denmark generates 40% of its electricity from wind, and at least 83 other countries around the world are using wind power to supply their electricity grids. In 2014 global wind power capacity expanded 16% to 369,553 MW. Yearly wind energy production is also growing rapidly and has reached around 4% of worldwide electricity usage, 11.4% in the EU.

History

Charles Brush's windmill of 1888, used for generating electricity.

Wind power has been used as long as humans have put sails into the wind. For more than two millennia wind-powered machines have ground grain and pumped water. Wind power was widely available and not confined to the banks of fast-flowing streams, or later, requiring sources of fuel. Wind-powered pumps drained the polders of the Netherlands, and in arid regions such as the American mid-west or the Australian outback, wind pumps provided water for live stock and steam engines.

The first windmill used for the production of electricity was built in Scotland in July 1887 by Prof James Blyth of Anderson's College, Glasgow (the precursor of Strathclyde University). Blyth's 10 m high, cloth-sailed wind turbine was installed in the garden of his holiday cottage at Marykirk in Kincardineshire and was used to charge accumulators developed by the Frenchman Camille Alphonse Faure, to power the lighting in the cottage, thus making it the first house in the world to have its electricity supplied by wind power. Blyth offered the surplus electricity to the people of Marykirk for lighting the main street, however, they turned down the offer as they thought electricity was "the work of the devil." Although he later built a wind turbine to supply emergency power to the local Lunatic Asylum, Infirmary and Dispensary of Montrose the invention never really caught on as the technology was not considered to be economically viable.

Across the Atlantic, in Cleveland, Ohio a larger and heavily engineered machine was designed and constructed in the winter of 1887–1888 by Charles F. Brush, this was built by his engineering company at his home and operated from 1886 until 1900. The Brush wind turbine had a rotor 17 m (56 foot) in diameter and was mounted on an 18 m (60 foot) tower. Although large by today's standards, the machine was only rated at 12 kW. The connected dynamo was used either to charge a bank of batteries or to operate up to 100 incandescent light bulbs, three arc lamps, and various motors in Brush's laboratory.

With the development of electric power, wind power found new applications in lighting buildings remote from centrally-generated power. Throughout the 20th century parallel paths developed small wind stations suitable for farms or residences, and larger utility-scale wind generators that could be connected to electricity grids for remote use of power. Today wind powered generators operate in every size range between tiny stations for battery charging at isolated residences, up to near-gigawatt sized offshore wind farms that provide electricity to national electrical networks.

Wind farms

Large onshore wind farms		
Wind farm	Current capacity (MW)	Country
Gansu Wind Farm	6,000	China
Muppandal wind farm	1,500	India
Alta (Oak Creek-Mojave)	1,320	United States
Jaisalmer Wind Park	1,064	India
Shepherds Flat Wind Farm	845	United States
Roscoe Wind Farm	782	United States
Horse Hollow Wind Energy Center	736	United States
Capricorn Ridge Wind Farm	662	United States
Fântânele-Cogealac Wind Farm	600	Romania
Fowler Ridge Wind Farm	600	United States
Whitelee Wind Farm	539	United Kingdom

A wind farm is a group of wind turbines in the same location used for production of electricity. A large wind farm may consist of several hundred individual wind turbines distributed over an extended area, but the land between the turbines may be used for agricultural or other purposes. For example, Gansu Wind Farm, the largest wind farm in the world, has several thousand turbines. A wind farm may also be located offshore.

Almost all large wind turbines have the same design — a horizontal axis wind turbine having an upwind rotor with three blades, attached to a nacelle on top of a tall tubular tower.

In a wind farm, individual turbines are interconnected with a medium voltage (often 34.5 kV), power collection system and communications network. In general, a distance of 7D (7 × Rotor Diameter of the Wind Turbine) is set between each turbine in a fully developed wind farm. At a substation, this medium-voltage electric current is increased in voltage with a transformer for connection to the high voltage electric power transmission system.

Generator Characteristics and Stability

Induction generators, which were often used for wind power projects in the 1980s and 1990s, require reactive power for excitation so substations used in wind-power collection systems include substantial capacitor banks for power factor correction. Different types of wind turbine generators behave differently during transmission grid disturbances, so extensive modelling of the dynamic electromechanical characteristics of a new wind farm is required by transmission system operators to ensure predictable stable behaviour during system faults. In particular, induction generators cannot support the system voltage during faults, unlike steam or hydro turbine-driven synchronous generators.

Today these generators aren't used any more in modern turbines. Instead today most turbines use variable speed generators combined with partial- or full-scale power converter between the turbine generator and the collector system, which generally have more desirable properties for grid interconnection and have Low voltage ride through-capabilities. Modern concepts use either doubly fed machines with partial-scale converters or squirrel-cage induction generators or synchronous generators (both permanently and electrically excited) with full scale converters.

Transmission systems operators will supply a wind farm developer with a grid code to specify the requirements for interconnection to the transmission grid. This will include power factor, constancy of frequency and dynamic behaviour of the wind farm turbines during a system fault.

Offshore Wind Power

The world's second full-scale floating wind turbine (and first to be installed without the use of heavy-lift vessels), WindFloat, operating at rated capacity (2 MW) approximately 5 km offshore of Póvoa de Varzim, Portugal

Offshore wind power refers to the construction of wind farms in large bodies of water to generate electricity. These installations can utilize the more frequent and powerful winds that are available in these locations and have less aesthetic impact on the landscape than land based projects. However, the construction and the maintenance costs are considerably higher.

Siemens and Vestas are the leading turbine suppliers for offshore wind power. DONG Energy, Vattenfall and E.ON are the leading offshore operators. As of October 2010, 3.16 GW of offshore wind power capacity was operational, mainly in Northern Europe. According to BTM Consult, more than 16 GW of additional capacity will be installed before the end of 2014 and the UK and Germany will become the two leading markets. Offshore wind power capacity is expected to reach a total of 75 GW worldwide by 2020, with significant contributions from China and the US.

In 2012, 1,662 turbines at 55 offshore wind farms in 10 European countries produced 18 TWh, enough to power almost five million households. As of August 2013 the London Array in the United Kingdom is the largest offshore wind farm in the world at 630 MW. This is followed by Gwynt y Môr (576 MW), also in the UK.

World's largest offshore wind farms				
Wind farm	**Capacity (MW)**	**Country**	**Turbines and model**	**Commissioned**
London Array	630	United Kingdom	175 × Siemens SWT-3.6	2012
Gwynt y Môr	576	United Kingdom	160 × Siemens SWT-3.6 107	2015
Greater Gabbard	504	United Kingdom	140 × Siemens SWT-3.6	2012
Anholt	400	Denmark	111 × Siemens SWT-3.6–120	2013
BARD Offshore 1	400	Germany	80 BARD 5.0 turbines	2013

Collection and Transmission Network

In a wind farm, individual turbines are interconnected with a medium voltage (usually 34.5 kV) power collection system and communications network. At a substation, this medium-voltage electric current is increased in voltage with a transformer for connection to the high voltage electric power transmission system.

A transmission line is required to bring the generated power to (often remote) markets. For an off-shore station this may require a submarine cable. Construction of a new high-voltage line may be too costly for the wind resource alone, but wind sites may take advantage of lines installed for conventionally fueled generation.

One of the biggest current challenges to wind power grid integration in the United States is the necessity of developing new transmission lines to carry power from wind farms, usually in remote lowly populated states in the middle of the country due to availability of wind, to high load locations, usually on the coasts where population density is higher. The current transmission lines in remote locations were not designed for the transport of large amounts of energy. As transmission lines become longer the losses associated with power transmission increase, as modes of losses at lower lengths are exacerbated and new modes of losses are no longer negligible as the length is increased, making it harder to transport large loads over large distances. However, resistance from state and local governments makes it difficult to construct new transmission lines. Multi state power transmission projects are discouraged by states with cheap electricity rates for fear that exporting their cheap power will lead to increased rates. A 2005 energy law gave the Energy Department authority to approve transmission projects states refused to act on, but after an attempt to use this authority, the Senate declared the department was being overly aggressive in doing so. Another problem is that wind companies find out after the fact that the transmission capacity of a new farm is below the generation capacity, largely because federal utility rules to encourage renewable energy installation allow feeder lines to meet only minimum standards. These are important issues that need to be solved, as when the transmission capacity does not meet the generation capacity, wind farms are forced to produce below their full potential or stop running all together, in a process known as curtailment. While this leads to potential renewable generation left untapped, it prevents possible grid overload or risk to reliable service.

Wind Power Capacity and Production

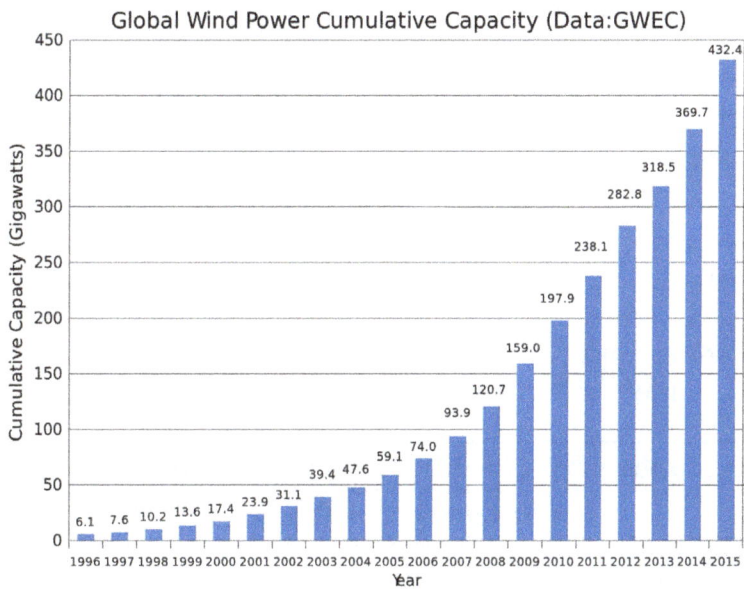

Global Wind Power Cumulative Capacity (Data:GWEC)

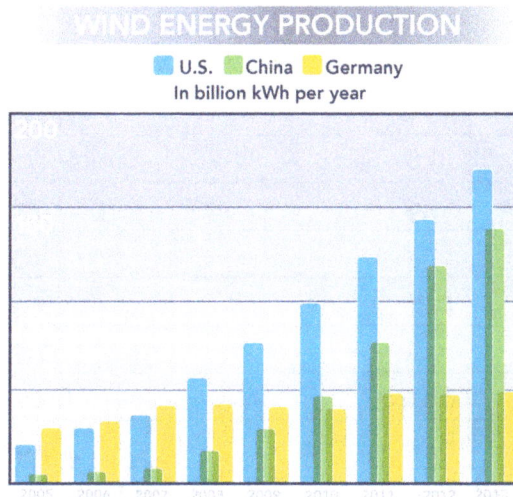

WIND ENERGY PRODUCTION

U.S. China Germany
In billion kWh per year

Worldwide there are now over two hundred thousand wind turbines operating, with a total nameplate capacity of 432,000 MW as of end 2015. The European Union alone passed some 100,000 MW nameplate capacity in September 2012, while the United States surpassed 75,000 MW in 2015 and China's grid connected capacity passed 145,000 MW in 2015.

World wind generation capacity more than quadrupled between 2000 and 2006, doubling about every three years. The United States pioneered wind farms and led the world in installed capacity in the 1980s and into the 1990s. In 1997 installed capacity in Germany surpassed the U.S. and led until once again overtaken by the U.S. in 2008. China has been rapidly expanding its wind instal-lations in the late 2000s and passed the U.S. in 2010 to become the world leader. As of 2011, 83 countries around the world were using wind power on a commercial basis.

Wind power capacity has expanded rapidly to 336 GW in June 2014, and wind energy production was around 4% of total worldwide electricity usage, and growing rapidly. The actual amount of electricity that wind is able to generate is calculated by multiplying the nameplate capacity by the capacity factor, which varies according to equipment and location. Estimates of the capacity factors for wind installations are in the range of 35% to 44%.

Europe accounted for 48% of the world total wind power generation capacity in 2009. In 2010, Spain became Europe's leading producer of wind energy, achieving 42,976 GWh. Germany held the top spot in Europe in terms of installed capacity, with a total of 27,215 MW as of 31 December 2010. In 2015 wind power constituted 15.6% of all installed power generation capacity in the EU and it generates around 11.4% of its power.

Top windpower electricity producing countries in 2012 (TWh)		
Country	**Windpower Production**	**% of World Total**
United States	140.9	26.4
China	118.1	22.1
Spain	49.1	9.2
Germany	46.0	8.6
India	30.0	5.6
United Kingdom	19.6	3.7
France	14.9	2.8
Italy	13.4	2.5
Canada	11.8	2.2
Denmark	10.3	1.9
(rest of world)	80.2	15.0
World Total	**534.3 TWh**	**100%**
Source:*Observ'ER – Electricity Production From Wind Sources*		

Growth Trends

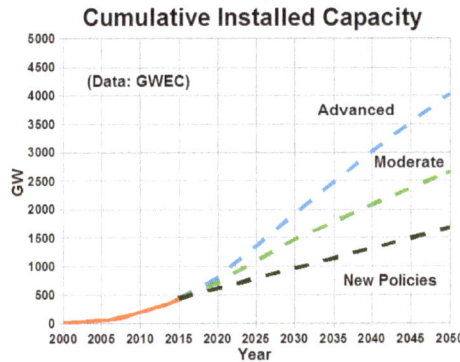

Cumulative Installed Capacity

(Data: GWEC)

Advanced

Moderate

New Policies

Worldwide installed wind power capacity forecast (Source: Global Wind Energy Council)

After setting new records in 2014, the wind power industry surprised many observers with another record breaking year in 2015, chalking up 22% annual market growth and passing the 60 GW mark for the first time in a single year; and this after having broken the 50 GW mark for the first time in 2014. In 2015, close to half of all new wind power was added outside of the traditional markets in Europe and North America. This was largely from new construction in China and India. Global Wind Energy Council (GWEC) figures show that 2015 recorded an increase of installed capacity of more than 63 GW, taking the total installed wind energy capacity to 432.9 GW, up from 74 GW in 2006. In terms of economic value, the wind energy sector has become one of the important players in the energy markets, with the total investments reaching US$329bn (€296.6bn), an increase of 4% over 2014.

Although the wind power industry was affected by the global financial crisis in 2009 and 2010, GWEC predicts that the installed capacity of wind power will be 792.1 GW by the end of 2020 and 4,042 GW by end of 2050. The increased commissioning of wind power is being accompanied by record low prices for forthcoming renewable electricity. In some cases, wind onshore is already the cheapest electricity generation option and costs are continuing to decline. The contracted prices for wind onshore for the next few years are now as low as 30 USD/MWh.

In the EU in 2015, 44% of all new generating capacity was wind power; while in the same period net fossil fuel power capacity decreased.

Capacity Factor

Since wind speed is not constant, a wind farm's annual energy production is never as much as the sum of the generator nameplate ratings multiplied by the total hours in a year. The ratio of actual productivity in a year to this theoretical maximum is called the capacity factor. Typical capacity factors are 15–50%; values at the upper end of the range are achieved in favourable sites and are due to wind turbine design improvements.

Online data is available for some locations, and the capacity factor can be calculated from the yearly output. For example, the German nationwide average wind power capacity factor over all of 2012 was just under 17.5% (45867 GW·h/yr / (29.9 GW × 24 × 366) = 0.1746), and the capacity factor for Scottish wind farms averaged 24% between 2008 and 2010.

Unlike fueled generating plants, the capacity factor is affected by several parameters, including the variability of the wind at the site and the size of the generator relative to the turbine's swept area. A small generator would be cheaper and achieve a higher capacity factor but would produce less electricity (and thus less profit) in high winds. Conversely, a large generator would cost more but generate little extra power and, depending on the type, may stall out at low wind speed. Thus an optimum capacity factor of around 40–50% would be aimed for.

A 2008 study released by the U.S. Department of Energy noted that the capacity factor of new wind installations was increasing as the technology improves, and projected further improvements for future capacity factors. In 2010, the department estimated the capacity factor of new wind turbines in 2010 to be 45%. The annual average capacity factor for wind generation in the US has varied between 29.8% and 34.0% during the period 2010–2015.

Penetration

Country	Penetration
Denmark (2015)	42.1%
Portugal (2011)	19%
Spain (2011)	16%
Ireland (2012)	16%
United Kingdom (2014)	9.3%
Germany (2011)	8%
United States (2013)	4.5%

Wind energy penetration is the fraction of energy produced by wind compared with the total generation. There is no generally accepted maximum level of wind penetration. The limit for a particular grid will depend on the existing generating plants, pricing mechanisms, capacity for energy storage, demand management and other factors. An interconnected electricity grid will already include reserve generating and transmission capacity to allow for equipment failures. This reserve capacity can also serve to compensate for the varying power generation produced by wind stations. Studies have indicated that 20% of the total annual electrical energy consumption may be incorporated with minimal difficulty. These studies have been for locations with geographically dispersed wind farms, some degree of dispatchable energy or hydropower with storage capacity, demand management, and interconnected to a large grid area enabling the export of electricity when needed. Beyond the 20% level, there are few technical limits, but the economic implications become more significant. Electrical utilities continue to study the effects of large scale penetration of wind generation on system stability and economics.

A wind energy penetration figure can be specified for different durations of time, but is often quoted annually. To obtain 100% from wind annually requires substantial long term storage or substantial interconnection to other systems which may already have substantial storage. On a monthly, weekly, daily, or hourly basis—or less—wind might supply as much as or more than 100% of current use, with the rest stored or exported. Seasonal industry might then take advantage of high wind and low usage times such as at night when wind output can exceed normal demand. Such industry might include production of silicon, aluminum, steel, or of natural gas, and hydrogen, and using future long term storage to facilitate 100% energy from variable renewable energy.

Homes can also be programmed to accept extra electricity on demand, for example by remotely turning up water heater thermostats.

In Australia, the state of South Australia generates around half of the nation's wind power capacity. By the end of 2011 wind power in South Australia, championed by Premier (and Climate Change Minister) Mike Rann, reached 26% of the State's electricity generation, edging out coal for the first time. At this stage South Australia, with only 7.2% of Australia's population, had 54% of Australia's installed capacity.

Variability

Windmills are typically installed in favourable windy locations. In the image,
wind power generators in Spain, near an Osborne bull.

Electricity generated from wind power can be highly variable at several different timescales: hourly, daily, or seasonally. Annual variation also exists, but is not as significant. Because instantaneous electrical generation and consumption must remain in balance to maintain grid stability, this variability can present substantial challenges to incorporating large amounts of wind power into a grid system. Intermittency and the non-dispatchable nature of wind energy production can raise costs for regulation, incremental operating reserve, and (at high penetration levels) could require an increase in the already existing energy demand management, load shedding, storage solutions or system interconnection with HVDC cables.

Fluctuations in load and allowance for failure of large fossil-fuel generating units require reserve capacity that can also compensate for variability of wind generation.

Increase in system operation costs, Euros per MWh, for 10% & 20% wind share		
Country	10%	20%
Germany	2.5	3.2
Denmark	0.4	0.8
Finland	0.3	1.5
Norway	0.1	0.3
Sweden	0.3	0.7

Wind power is variable, and during low wind periods it must be replaced by other power sources. Transmission networks presently cope with outages of other generation plants and daily changes in electrical demand, but the variability of intermittent power sources such as wind power, are unlike those of conventional power generation plants which, when scheduled to be operating, may be able to deliver their nameplate capacity around 95% of the time.

Presently, grid systems with large wind penetration require a small increase in the frequency of usage of natural gas spinning reserve power plants to prevent a loss of electricity in the event that conditions are not favorable for power production from the wind. At lower wind power grid penetration, this is less of an issue.

GE has installed a prototype wind turbine with onboard battery similar to that of an electric car, equivalent of 1 minute of production. Despite the small capacity, it is enough to guarantee that power output complies with forecast for 15 minutes, as the battery is used to eliminate the difference rather than provide full output. The increased predictability can be used to take wind power penetration from 20 to 30 or 40 per cent. The battery cost can be retrieved by selling burst power on demand and reducing backup needs from gas plants.

A report on Denmark's wind power noted that their wind power network provided less than 1% of average demand on 54 days during the year 2002. Wind power advocates argue that these periods of low wind can be dealt with by simply restarting existing power stations that have been held in readiness, or interlinking with HVDC. Electrical grids with slow-responding thermal power plants and without ties to networks with hydroelectric generation may have to limit the use of wind power. According to a 2007 Stanford University study published in the *Journal of Applied Meteorology and Climatology*, interconnecting ten or more wind farms can allow an average of 33% of the total energy produced (i.e. about 8% of total nameplate capacity) to be used as reliable, baseload electric power which can be relied on to handle peak loads, as long as minimum criteria are met for wind speed and turbine height.

Conversely, on particularly windy days, even with penetration levels of 16%, wind power generation can surpass all other electricity sources in a country. In Spain, in the early hours of 16 April 2012 wind power production reached the highest percentage of electricity production till then, at 60.46% of the total demand. In Denmark, which had power market penetration of 30% in 2013, over 90 hours, wind power generated 100% of the country's power, peaking at 122% of the country's demand at 2 am on 28 October.

A 2006 International Energy Agency forum presented costs for managing intermittency as a function of wind-energy's share of total capacity for several countries, as shown in the table on the right. Three reports on the wind variability in the UK issued in 2009, generally agree that variability of wind needs to be taken into account, but it does not make the grid unmanageable. The additional costs, which are modest, can be quantified.

The combination of diversifying variable renewables by type and location, forecasting their variation, and integrating them with dispatchable renewables, flexible fueled generators, and demand response can create a power system that has the potential to meet power supply needs reliably. Integrating ever-higher levels of renewables is being successfully demonstrated in the real world:

In 2009, eight American and three European authorities, writing in the leading electrical engineers' professional journal, didn't find "a credible and firm technical limit to the amount of wind energy that can be accommodated by electricity grids". In fact, not one of more than 200 international studies, nor official studies for the eastern and western U.S. regions, nor the International Energy Agency, has found major costs or technical barriers to reliably integrating up to 30% variable renewable supplies into the grid, and in some studies much more. – *Reinventing Fire*

Solar power tends to be complementary to wind. On daily to weekly timescales, high pressure areas tend to bring clear skies and low surface winds, whereas low pressure areas tend to be windier and cloudier. On seasonal timescales, solar energy peaks in summer, whereas in many areas wind energy is lower in summer and higher in winter.[nb 2] Thus the intermittencies of wind and solar power tend to cancel each other somewhat. In 2007 the Institute for Solar Energy Supply Technology of the University of Kassel pilot-tested a combined power plant linking solar, wind, biogas and hydrostorage to provide load-following power around the clock and throughout the year, entirely from renewable sources.

Predictability

Wind power forecasting methods are used, but predictability of any particular wind farm is low for short-term operation. For any particular generator there is an 80% chance that wind output will change less than 10% in an hour and a 40% chance that it will change 10% or more in 5 hours.

However, studies by Graham Sinden (2009) suggest that, in practice, the variations in thousands of wind turbines, spread out over several different sites and wind regimes, are smoothed. As the distance between sites increases, the correlation between wind speeds measured at those sites, decreases.

Thus, while the output from a single turbine can vary greatly and rapidly as local wind speeds vary, as more turbines are connected over larger and larger areas the average power output becomes less variable and more predictable.

Wind power hardly ever suffers major technical failures, since failures of individual wind turbines have hardly any effect on overall power, so that the distributed wind power is reliable and predictable, whereas conventional generators, while far less variable, can suffer major unpredictable outages.

Energy Storage

The Sir Adam Beck Generating Complex at Niagara Falls, Canada, includes a large pumped-storage hydroelectricity reservoir. During hours of low electrical demand excess electrical grid power is used to pump water up into the reservoir, which then provides an extra 174 MW of electricity during periods of peak demand.

Typically, conventional hydroelectricity complements wind power very well. When the wind is blowing strongly, nearby hydroelectric stations can temporarily hold back their water. When the wind drops they can, provided they have the generation capacity, rapidly increase production to compensate. This gives a very even overall power supply and virtually no loss of energy and uses no more water.

Alternatively, where a suitable head of water is not available, pumped-storage hydroelectricity or other forms of grid energy storage such as compressed air energy storage and thermal energy storage can store energy developed by high-wind periods and release it when needed. The type of storage needed depends on the wind penetration level – low penetration requires daily storage, and high penetration requires both short and long term storage – as long as a month or more. Stored energy increases the economic value of wind energy since it can be shifted to displace higher cost generation during peak demand periods. The potential revenue from this arbitrage can offset the cost and losses of storage; the cost of storage may add 25% to the cost of any wind energy stored but it is not envisaged that this would apply to a large proportion of wind energy generated. For example, in the UK, the 1.7 GW Dinorwig pumped-storage plant evens out electrical demand peaks, and allows base-load suppliers to run their plants more efficiently. Although pumped-storage power systems are only about 75% efficient, and have high installation costs, their low running costs and ability to reduce the required electrical base-load can save both fuel and total electrical generation costs.

In particular geographic regions, peak wind speeds may not coincide with peak demand for electrical power. In the U.S. states of California and Texas, for example, hot days in summer may have low wind speed and high electrical demand due to the use of air conditioning. Some utilities subsidize the purchase of geothermal heat pumps by their customers, to reduce electricity demand during the summer months by making air conditioning up to 70% more efficient; widespread adoption of this technology would better match electricity demand to wind availability in areas with hot summers and low summer winds. A possible future option may be to interconnect widely dispersed geographic areas with an HVDC "super grid". In the U.S. it is estimated that to upgrade the transmission system to take in planned or potential renewables would cost at least $60 billion, while the society value of added windpower would be more than that cost.

Germany has an installed capacity of wind and solar that can exceed daily demand, and has been exporting peak power to neighboring countries, with exports which amounted to some 14.7 billion kilowatt hours in 2012. A more practical solution is the installation of thirty days storage capacity able to supply 80% of demand, which will become necessary when most of Europe's energy is obtained from wind power and solar power. Just as the EU requires member countries to maintain 90 days strategic reserves of oil it can be expected that countries will provide electricity storage, instead of expecting to use their neighbors for net metering.

Capacity Credit, Fuel Savings and Energy Payback

The capacity credit of wind is estimated by determining the capacity of conventional plants displaced by wind power, whilst maintaining the same degree of system security. According to the American Wind Energy Association, production of wind power in the United States in 2015 avoided consumption of 73 billion gallons of water and reduced CO_2 emissions by 132 million metric tons, while providing $7.3 billion in public health savings.

The energy needed to build a wind farm divided into the total output over its life, Energy Return on Energy Invested, of wind power varies but averages about 20–25. Thus, the energy payback time is typically around one year.

Economics

Wind turbines reached grid parity (the point at which the cost of wind power matches traditional sources) in some areas of Europe in the mid-2000s, and in the US around the same time. Falling prices continue to drive the levelized cost down and it has been suggested that it has reached general grid parity in Europe in 2010, and will reach the same point in the US around 2016 due to an expected reduction in capital costs of about 12%.

Electricity cost and trends

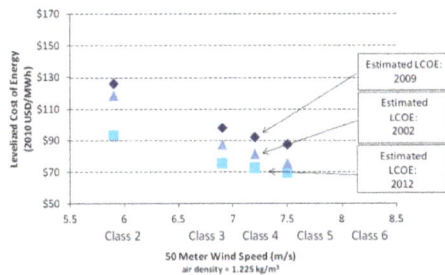

Estimated cost per MWh for wind power in Denmark

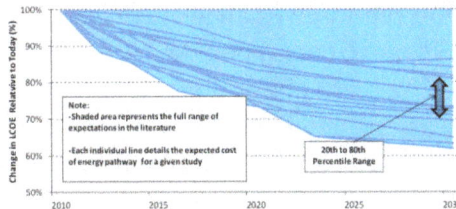

The National Renewable Energy Laboratory projects that the levelized cost of wind power in the U.S. will decline about 25% from 2012 to 2030.

A turbine blade convoy passing through Edenfield in the U.K. (2008). Even longer two-piece blades are now manufactured, and then assembled on-site to reduce difficulties in transportation.

Wind power is capital intensive, but has no fuel costs. The price of wind power is therefore much more stable than the volatile prices of fossil fuel sources. The marginal cost of wind energy once a station is constructed is usually less than 1-cent per kW·h.

However, the estimated average cost per unit of electricity must incorporate the cost of construction of the turbine and transmission facilities, borrowed funds, return to investors (including cost of risk), estimated annual production, and other components, averaged over the projected useful life of the equipment, which may be in excess of twenty years. Energy cost estimates are highly dependent on these assumptions so published cost figures can differ substantially. In 2004, wind energy cost a fifth of what it did in the 1980s, and some expected that downward trend to continue as larger multi-megawatt turbines were mass-produced. In 2012 capital costs for wind turbines were substantially lower than 2008–2010 but still above 2002 levels. A 2011 report from the American Wind Energy Association stated, "Wind's costs have dropped over the past two years, in the range of 5 to 6 cents per kilowatt-hour recently.... about 2 cents cheaper than coal-fired electricity, and more projects were financed through debt arrangements than tax equity structures last year.... winning more mainstream acceptance from Wall Street's banks.... Equipment makers can also deliver products in the same year that they are ordered instead of waiting up to three years as was the case in previous cycles.... 5,600 MW of new installed capacity is under construction in the United States, more than double the number at this point in 2010. Thirty-five percent of all new power generation built in the United States since 2005 has come from wind, more than new gas and coal plants combined, as power providers are increasingly enticed to wind as a convenient hedge against unpredictable commodity price moves."

A British Wind Energy Association report gives an average generation cost of onshore wind power of around 3.2 pence (between US 5 and 6 cents) per kW·h (2005). Cost per unit of energy produced was estimated in 2006 to be 5 to 6 percent above the cost of new generating capacity in the US for coal and natural gas: wind cost was estimated at $55.80 per MW·h, coal at $53.10/MW·h and natural gas at $52.50. Similar comparative results with natural gas were obtained in a governmental study in the UK in 2011. In 2011 power from wind turbines could be already cheaper than fossil or nuclear plants; it is also expected that wind power will be the cheapest form of energy generation in the future. The presence of wind energy, even when subsidised, can reduce costs for consumers (€5 billion/yr in Germany) by reducing the marginal price, by minimising the use of expensive peaking power plants.

An 2012 EU study shows base cost of onshore wind power similar to coal, when subsidies and externalities are disregarded. Wind power has some of the lowest external costs.

In February 2013 Bloomberg New Energy Finance (BNEF) reported that the cost of generating electricity from new wind farms is cheaper than new coal or new baseload gas plants. When including the current Australian federal government carbon pricing scheme their modeling gives costs (in Australian dollars) of $80/MWh for new wind farms, $143/MWh for new coal plants and $116/MWh for new baseload gas plants. The modeling also shows that "even without a carbon price (the most efficient way to reduce economy-wide emissions) wind energy is 14% cheaper than new coal and 18% cheaper than new gas." Part of the higher costs for new coal plants is due to high financial lending costs because of "the reputational damage of emissions-intensive investments". The expense of gas fired plants is partly due to "export market" effects on local prices. Costs of production from coal fired plants built in "the 1970s and 1980s" are cheaper than renewable energy sources because of depreciation. In 2015 BNEF calculated LCOE prices per MWh energy in new power-plants (excluding carbon costs) : $85 for onshore wind ($175 for offshore), $66–75 for coal in the Americas ($82–105 in Europe), gas $80–100. A 2014 study showed unsubsidized LCOE costs between $37–81, depending on region. A 2014 US DOE report showed that in some cases power purchase agreement prices for wind power had dropped to record lows of $23.5/MWh.

The cost has reduced as wind turbine technology has improved. There are now longer and lighter wind turbine blades, improvements in turbine performance and increased power generation efficiency. Also, wind project capital and maintenance costs have continued to decline. For example, the wind industry in the USA in early 2014 were able to produce more power at lower cost by using taller wind turbines with longer blades, capturing the faster winds at higher elevations. This has opened up new opportunities and in Indiana, Michigan, and Ohio, the price of power from wind turbines built 300 feet to 400 feet above the ground can now compete with conventional fossil fuels like coal. Prices have fallen to about 4 cents per kilowatt-hour in some cases and utilities have been increasing the amount of wind energy in their portfolio, saying it is their cheapest option.

A number of initiatives are working to reduce costs of electricity from offshore wind. One example is the Carbon Trust Offshore Wind Accelerator, a joint industry project, involving nine offshore wind developers, which aims to reduce the cost of offshore wind by 10% by 2015. It has been suggested that innovation at scale could deliver 25% cost reduction in offshore wind by 2020. Henrik Stiesdal, former Chief Technical Officer at Siemens Wind Power, has stated that by 2025 energy from offshore wind will be one of the cheapest, scalable solutions in the UK, compared to other renewables and fossil fuel energy sources, if the true cost to society is factored into the cost of energy equation. He calculates the cost at that time to be 43 EUR/MWh for onshore, and 72 EUR/MWh for offshore wind.

Incentives and Community Benefits

U.S. landowners typically receive $3,000–$5,000 annual rental income per wind turbine, while farmers continue to grow crops or graze cattle up to the foot of the turbines. Shown: the Brazos Wind Farm, Texas.

Some of the 6,000 turbines in California's Altamont Pass Wind Farm aided by tax incentives during the 1980s.

The U.S. wind industry generates tens of thousands of jobs and billions of dollars of economic activity. Wind projects provide local taxes, or payments in lieu of taxes and strengthen the economy of rural communities by providing income to farmers with wind turbines on their land. Wind energy in many jurisdictions receives financial or other support to encourage its development. Wind energy benefits from subsidies in many jurisdictions, either to increase its attractiveness, or to compensate for subsidies received by other forms of production which have significant negative externalities.

In the US, wind power receives a production tax credit (PTC) of 1.5¢/kWh in 1993 dollars for each kW·h produced, for the first ten years; at 2.2 cents per kW·h in 2012, the credit was renewed on 2 January 2012, to include construction begun in 2013. A 30% tax credit can be applied instead of receiving the PTC. Another tax benefit is accelerated depreciation. Many American states also provide incentives, such as exemption from property tax, mandated purchases, and additional markets for "green credits". The Energy Improvement and Extension Act of 2008 contains extensions of credits for wind, including microturbines. Countries such as Canada and Germany also provide incentives for wind turbine construction, such as tax credits or minimum purchase prices for wind generation, with assured grid access (sometimes referred to as feed-in tariffs). These feed-in tariffs are typically set well above average electricity prices. In December 2013 U.S. Senator Lamar Alexander and other Republican senators argued that the "wind energy production tax credit should be allowed to expire at the end of 2013" and it expired 1 January 2014 for new installations.

Secondary market forces also provide incentives for businesses to use wind-generated power, even if there is a premium price for the electricity. For example, socially responsible manufacturers pay utility companies a premium that goes to subsidize and build new wind power infrastructure. Companies use wind-generated power, and in return they can claim that they are undertaking strong "green" efforts. In the US the organization Green-e monitors business compliance with these renewable energy credits.

Small-scale Wind Power

A small Quietrevolution QR5 Gorlov type vertical axis wind turbine on the roof of Colston Hall in Bristol, England. Measuring 3 m in diameter and 5 m high, it has a nameplate rating of 6.5 kW.

Small-scale wind power is the name given to wind generation systems with the capacity to produce up to 50 kW of electrical power. Isolated communities, that may otherwise rely on diesel generators, may use wind turbines as an alternative. Individuals may purchase these systems to reduce or eliminate their dependence on grid electricity for economic reasons, or to reduce their carbon footprint. Wind turbines have been used for household electricity generation in conjunction with battery storage over many decades in remote areas.

Recent examples of small-scale wind power projects in an urban setting can be found in New York City, where, since 2009, a number of building projects have capped their roofs with Gorlov-type helical wind turbines. Although the energy they generate is small compared to the buildings' overall consumption, they help to reinforce the building's 'green' credentials in ways that "showing people your high-tech boiler" can not, with some of the projects also receiving the direct support of the New York State Energy Research and Development Authority.

Grid-connected domestic wind turbines may use grid energy storage, thus replacing purchased electricity with locally produced power when available. The surplus power produced by domestic microgenerators can, in some jurisdictions, be fed into the network and sold to the utility company, producing a retail credit for the microgenerators' owners to offset their energy costs.

Off-grid system users can either adapt to intermittent power or use batteries, photovoltaic or diesel systems to supplement the wind turbine. Equipment such as parking meters, traffic warning signs, street lighting, or wireless Internet gateways may be powered by a small wind turbine, possibly combined with a photovoltaic system, that charges a small battery replacing the need for a connection to the power grid.

A Carbon Trust study into the potential of small-scale wind energy in the UK, published in 2010, found that small wind turbines could provide up to 1.5 terawatt hours (TW·h) per year of electricity

(0.4% of total UK electricity consumption), saving 0.6 million tonnes of carbon dioxide (Mt CO_2) emission savings. This is based on the assumption that 10% of households would install turbines at costs competitive with grid electricity, around 12 pence (US 19 cents) a kW·h. A report prepared for the UK's government-sponsored Energy Saving Trust in 2006, found that home power generators of various kinds could provide 30 to 40% of the country's electricity needs by 2050.

Distributed generation from renewable resources is increasing as a consequence of the increased awareness of climate change. The electronic interfaces required to connect renewable generation units with the utility system can include additional functions, such as the active filtering to enhance the power quality.

Environmental Effects

Livestock grazing near a wind turbine.

The environmental impact of wind power when compared to the environmental impacts of fossil fuels, is relatively minor. According to the IPCC, in assessments of the life-cycle global warming potential of energy sources, wind turbines have a median value of between 12 and 11 (gCO_{2eq}/kWh) depending on whether off- or onshore turbines are being assessed. Compared with other low carbon power sources, wind turbines have some of the lowest global warming potential per unit of electrical energy generated.

While a wind farm may cover a large area of land, many land uses such as agriculture are compatible with it, as only small areas of turbine foundations and infrastructure are made unavailable for use.

There are reports of bird and bat mortality at wind turbines as there are around other artificial structures. The scale of the ecological impact may or may not be significant, depending on specific circumstances. Prevention and mitigation of wildlife fatalities, and protection of peat bogs, affect the siting and operation of wind turbines.

Wind turbines generate some noise. At a residential distance of 300 metres (980 ft) this may be around 45 dB, which is slightly louder than a refrigerator. At 1.5 km (1 mi) distance they become inaudible. There are anecdotal reports of negative health effects from noise on people who live very close to wind turbines. Peer-reviewed research has generally not supported these claims.

Aesthetic aspects of wind turbines and resulting changes of the visual landscape are significant. Conflicts arise especially in scenic and heritage protected landscapes.

Politics

Central Government

Part of the Seto Hill Windfarm in Japan.

Nuclear power and fossil fuels are subsidized by many governments, and wind power and other forms of renewable energy are also often subsidized. For example, a 2009 study by the Environmental Law Institute assessed the size and structure of U.S. energy subsidies over the 2002–2008 period. The study estimated that subsidies to fossil-fuel based sources amounted to approximately $72 billion over this period and subsidies to renewable fuel sources totalled $29 billion. In the United States, the federal government has paid US$74 billion for energy subsidies to support R&D for nuclear power ($50 billion) and fossil fuels ($24 billion) from 1973 to 2003. During this same time frame, renewable energy technologies and energy efficiency received a total of US$26 billion. It has been suggested that a subsidy shift would help to level the playing field and support growing energy sectors, namely solar power, wind power, and biofuels. History shows that no energy sector was developed without subsidies.

According to the International Energy Agency (IEA) (2011), energy subsidies artificially lower the price of energy paid by consumers, raise the price received by producers or lower the cost of production. "Fossil fuels subsidies costs generally outweigh the benefits. Subsidies to renewables and low-carbon energy technologies can bring long-term economic and environmental benefits". In November 2011, an IEA report entitled *Deploying Renewables 2011* said "subsidies in green energy technologies that were not yet competitive are justified in order to give an incentive to investing into technologies with clear environmental and energy security benefits". The IEA's report disagreed with claims that renewable energy technologies are only viable through costly subsidies and not able to produce energy reliably to meet demand.

In the U.S., the wind power industry has recently increased its lobbying efforts considerably, spending about $5 million in 2009 after years of relative obscurity in Washington. By comparison, the U.S. nuclear industry alone spent over $650 million on its lobbying efforts and campaign contributions during a single ten-year period ending in 2008.

Following the 2011 Japanese nuclear accidents, Germany's federal government is working on a new plan for increasing energy efficiency and renewable energy commercialization, with a particular focus on offshore wind farms. Under the plan, large wind turbines will be erected far away from the coastlines, where the wind blows more consistently than it does on land, and where the enormous turbines won't bother the inhabitants. The plan aims to decrease Germany's dependence on energy derived from coal and nuclear power plants.

Public Opinion

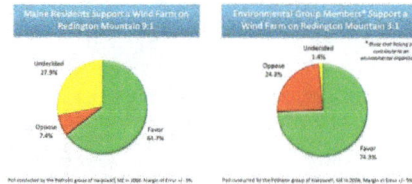

Environmental group members are both more in favor of wind power (74%) as well as more opposed (24%). Few are undecided.

Surveys of public attitudes across Europe and in many other countries show strong public support for wind power. About 80% of EU citizens support wind power. In Germany, where wind power has gained very high social acceptance, hundreds of thousands of people have invested in citizens' wind farms across the country and thousands of small and medium-sized enterprises are running successful businesses in a new sector that in 2008 employed 90,000 people and generated 8% of Germany's electricity.

Although wind power is a popular form of energy generation, the construction of wind farms is not universally welcomed, often for aesthetic reasons.

In Spain, with some exceptions, there has been little opposition to the installation of inland wind parks. However, the projects to build offshore parks have been more controversial. In particular, the proposal of building the biggest offshore wind power production facility in the world in southwestern Spain in the coast of Cádiz, on the spot of the 1805 Battle of Trafalgar has been met with strong opposition who fear for tourism and fisheries in the area, and because the area is a war grave.

In a survey conducted by Angus Reid Strategies in October 2007, 89 per cent of respondents said that using renewable energy sources like wind or solar power was positive for Canada, because these sources were better for the environment. Only 4 per cent considered using renewable sources as negative since they can be unreliable and expensive. According to a Saint Consulting survey in April 2007, wind power was the alternative energy source most likely to gain public support for future development in Canada, with only 16% opposed to this type of energy. By contrast, 3 out of 4 Canadians opposed nuclear power developments.

A 2003 survey of residents living around Scotland's 10 existing wind farms found high levels of community acceptance and strong support for wind power, with much support from those who lived closest to the wind farms. The results of this survey support those of an earlier Scottish Executive survey 'Public attitudes to the Environment in Scotland 2002', which found that the Scottish public would prefer the majority of their electricity to come from renewables, and which rated wind power as the cleanest source of renewable energy. A survey conducted in 2005 showed that 74% of people in Scotland agree that wind farms are necessary to meet current and future energy needs. When people were asked the same question in a Scottish renewables study conducted in 2010, 78% agreed. The increase is significant as there were twice as many wind farms in 2010 as there were in 2005. The 2010 survey also showed that 52% disagreed with the statement that wind farms are "ugly and a blot on the landscape". 59% agreed that wind farms were necessary and that how they looked was unimportant. Regarding tourism, query responders consider power pylons, cell phone towers, quarries and plantations more negatively than wind farms. Scotland is planning to obtain 100% of electricity from renewable sources by 2020.

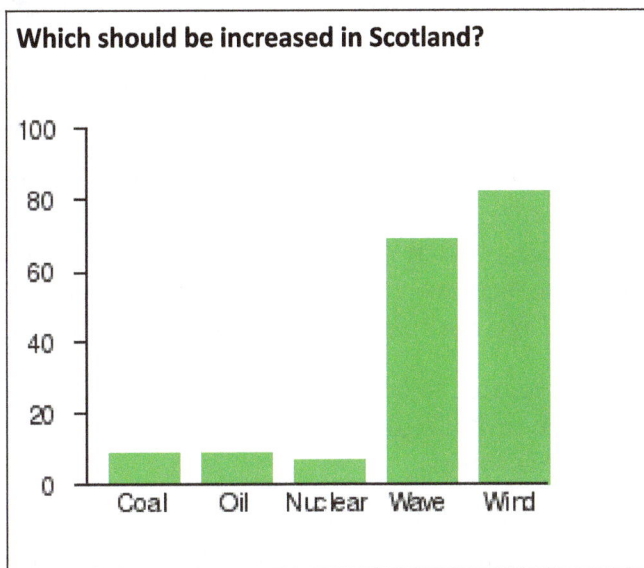

In other cases there is direct community ownership of wind farm projects. The hundreds of thousands of people who have become involved in Germany's small and medium-sized wind farms demonstrate such support there.

This 2010 Harris Poll reflects the strong support for wind power in Germany, other European countries, and the U.S.

Opinion on increase in number of wind farms, 2010 Harris Poll						
	U.S.	Great Britain	France	Italy	Spain	Germany
	%	%	%	%	%	%
Strongly oppose	3	6	6	2	2	4
Oppose more than favour	9	12	16	11	9	14
Favour more than oppose	37	44	44	38	37	42
Strongly favour	50	38	33	49	53	40

Community

Wind turbines such as these, in Cumbria, England, have been opposed for a number of reasons,
including aesthetics, by some sectors of the population.

Many wind power companies work with local communities to reduce environmental and other
concerns associated with particular wind farms. In other cases there is direct community ownership of wind farm projects. Appropriate government consultation, planning and approval procedures also help to minimize environmental risks. Some may still object to wind farms but, according to The Australia Institute, their concerns should be weighed against the need to address the
threats posed by climate change and the opinions of the broader community.

In America, wind projects are reported to boost local tax bases, helping to pay for schools, roads
and hospitals. Wind projects also revitalize the economy of rural communities by providing steady
income to farmers and other landowners.

In the UK, both the National Trust and the Campaign to Protect Rural England have expressed
concerns about the effects on the rural landscape caused by inappropriately sited wind turbines
and wind farms.

A panoramic view of the United Kingdom's Whitelee Wind Farm with Lochgoin Reservoir in the foreground.

Some wind farms have become tourist attractions. The Whitelee Wind Farm Visitor Centre has
an exhibition room, a learning hub, a café with a viewing deck and also a shop. It is run by the
Glasgow Science Centre.

In Denmark, a loss-of-value scheme gives people the right to claim compensation for loss of value of their property if it is caused by proximity to a wind turbine. The loss must be at least 1% of the property's value.

Despite this general support for the concept of wind power in the public at large, local opposition often exists and has delayed or aborted a number of projects. For example, there are concerns that some installations can negatively affect TV and radio reception and Doppler weather radar, as well as produce excessive sound and vibration levels leading to a decrease in property values. Potential broadcast-reception solutions include predictive interference modeling as a component of site selection. A study of 50,000 home sales near wind turbines found no statistical evidence that prices were affected.

While aesthetic issues are subjective and some find wind farms pleasant and optimistic, or symbols of energy independence and local prosperity, protest groups are often formed to attempt to block new wind power sites for various reasons.

This type of opposition is often described as NIMBYism, but research carried out in 2009 found that there is little evidence to support the belief that residents only object to renewable power facilities such as wind turbines as a result of a "Not in my Back Yard" attitude.

Turbine Design

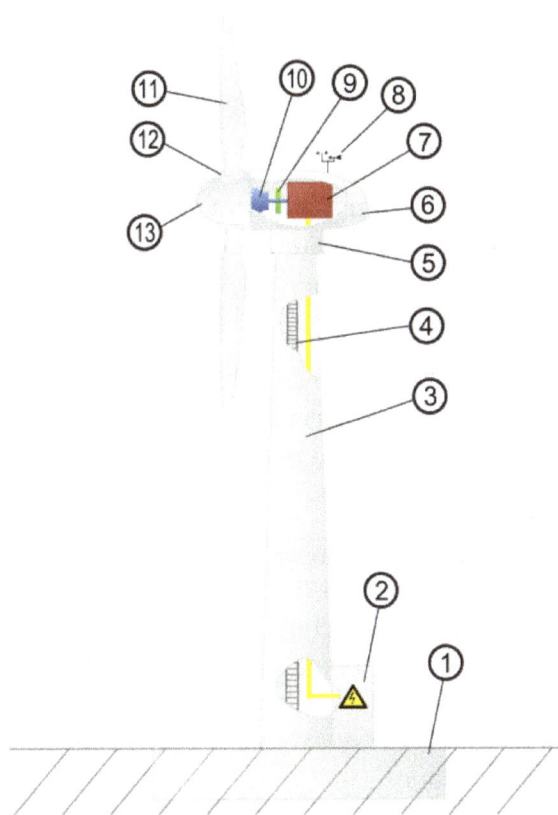

Typical wind turbine components : 1-Foundation, 2-Connection to the electric grid, 3-Tower, 4-Access ladder, 5-Wind orientation control (Yaw control), 6-Nacelle, 7-Generator, 8-Anemometer, 9-Electric or Mechanical Brake, 10-Gearbox, 11-Rotor blade, 12-Blade pitch control, 13-Rotor hub.

Typical components of a wind turbine (gearbox, rotor shaft and brake assembly) being lifted into position

Wind turbines are devices that convert the wind's kinetic energy into electrical power. The result of over a millennium of windmill development and modern engineering, today's wind turbines are manufactured in a wide range of horizontal axis and vertical axis types. The smallest turbines are used for applications such as battery charging for auxiliary power. Slightly larger turbines can be used for making small contributions to a domestic power supply while selling unused power back to the utility supplier via the electrical grid. Arrays of large turbines, known as wind farms, have become an increasingly important source of renewable energy and are used in many countries as part of a strategy to reduce their reliance on fossil fuels.

Wind turbine design is the process of defining the form and specifications of a wind turbine to extract energy from the wind. A wind turbine installation consists of the necessary systems needed to capture the wind's energy, point the turbine into the wind, convert mechanical rotation into electrical power, and other systems to start, stop, and control the turbine.

In 1919 the German physicist Albert Betz showed that for a hypothetical ideal wind-energy extraction machine, the fundamental laws of conservation of mass and energy allowed no more than 16/27 (59.3%) of the kinetic energy of the wind to be captured. This Betz limit can be approached in modern turbine designs, which may reach 70 to 80% of the theoretical Betz limit.

The aerodynamics of a wind turbine are not straightforward. The air flow at the blades is not the same as the airflow far away from the turbine. The very nature of the way in which energy is extracted from the air also causes air to be deflected by the turbine. In addition the aerodynamics of a wind turbine at the rotor surface exhibit phenomena that are rarely seen in other aerodynamic fields. The shape and dimensions of the blades of the wind turbine are determined by the aerodynamic performance required to efficiently extract energy from the wind, and by the strength required to resist the forces on the blade.

In addition to the aerodynamic design of the blades, the design of a complete wind power system must also address the design of the installation's rotor hub, nacelle, tower structure, generator, controls, and foundation. Further design factors must also be considered when integrating wind turbines into electrical power grids.

Wind Energy

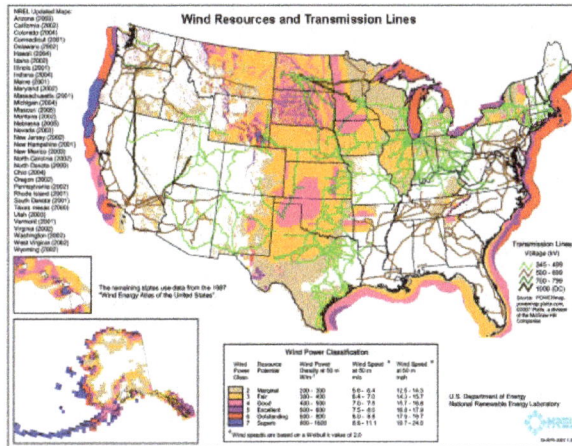

Map of available wind power for the United States. Color codes indicate wind power density class.

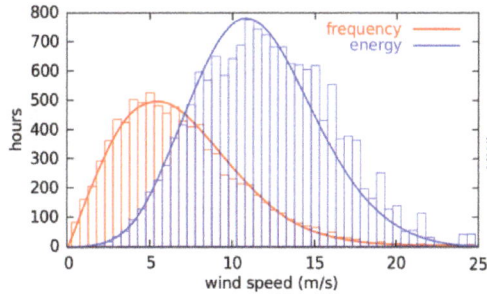

Distribution of wind speed (red) and energy (blue) for all of 2002 at the Lee Ranch facility in Colorado. The histogram shows measured data, while the curve is the Rayleigh model distribution for the same average wind speed.

Wind energy is the kinetic energy of air in motion, also called wind. Total wind energy flowing through an imaginary surface with area A during the time t is:

$$E = \frac{1}{2}mv^2 = \frac{1}{2}(Avt\rho)v^2 = \frac{1}{2}At\rho v^3,$$

where ρ is the density of air; v is the wind speed; Avt is the volume of air passing through A (which is considered perpendicular to the direction of the wind); $Avt\rho$ is therefore the mass m passing through "A". Note that ½ ρv^2 is the kinetic energy of the moving air per unit volume.

Power is energy per unit time, so the wind power incident on A (e.g. equal to the rotor area of a wind turbine) is:

$$P = \frac{E}{t} = \frac{1}{2}A\rho v^3$$

Wind power in an open air stream is thus *proportional* to the *third power* of the wind speed; the available power increases eightfold when the wind speed doubles. Wind turbines for grid electricity therefore need to be especially efficient at greater wind speeds.

Wind is the movement of air across the surface of the Earth, affected by areas of high pressure and of low pressure. The global wind kinetic energy averaged approximately 1.50 MJ/m² over the period from 1979 to 2010, 1.31 MJ/m² in the Northern Hemisphere with 1.70 MJ/m² in the Southern Hemisphere. The atmosphere acts as a thermal engine, absorbing heat at higher temperatures, releasing heat at lower temperatures. The process is responsible for production of wind kinetic energy at a rate of 2.46 W/m² sustaining thus the circulation of the atmosphere against frictional dissipation. A global 1 km² map of wind resources is housed at http://irena.masdar.ac.ae/?map=103, based on calculations by the Technical University of Denmark.

The total amount of economically extractable power available from the wind is considerably more than present human power use from all sources. Axel Kleidon of the Max Planck Institute in Germany, carried out a "top down" calculation on how much wind energy there is, starting with the incoming solar radiation that drives the winds by creating temperature differences in the atmosphere. He concluded that somewhere between 18 TW and 68 TW could be extracted.

Cristina Archer and Mark Z. Jacobson presented a "bottom-up" estimate, which unlike Kleidon's are based on actual measurements of wind speeds, and found that there is 1700 TW of wind power at an altitude of 100 metres over land and sea. Of this, "between 72 and 170 TW could be extracted in a practical and cost-competitive manner". They later estimated 80 TW. However research at Harvard University estimates 1 Watt/m² on average and 2–10 MW/km² capacity for large scale wind farms, suggesting that these estimates of total global wind resources are too high by a factor of about 4.

The strength of wind varies, and an average value for a given location does not alone indicate the amount of energy a wind turbine could produce there.

To assess prospective wind power sites a probability distribution function is often fit to the observed wind speed data. Different locations will have different wind speed distributions. The Weibull model closely mirrors the actual distribution of hourly/ten-minute wind speeds at many locations. The Weibull factor is often close to 2 and therefore a Rayleigh distribution can be used as a less accurate, but simpler model.

Gallery

REpower 5 MW wind turbine under construction at Nigg fabrication yard on the Cromarty Firth

The London Array under construction in 2009

Sunrise at the Fenton Wind Farm in Minnesota, United States.

Wind farm in Xinjiang, China

Scroby Sands wind farm from Great Yarmouth

A wind turbine blade on I-35 near Elm Mott, an increasingly common sight in Texas

Erection of an Enercon E70-4 in Germany.

Middelgrunden offshore wind park.

Bozcaada Wind Farm in Çanakkale province, Turkey.

Wave Power

Pelamis Wave Energy Converter on site at the European Marine Energy Centre (EMEC), in 2008

Azura at the US Navy's Wave Energy Test Site (WETS) on Oahu

SINN Power Wave Energy Converter

Wave power is the transport of energy by wind waves, and the capture of that energy to do useful work – for example, electricity generation, water desalination, or the pumping of water (into reservoirs). A machine able to exploit wave power is generally known as a wave energy converter (WEC).

Wave power is distinct from the diurnal flux of tidal power and the steady gyre of ocean currents. Wave-power generation is not currently a widely employed commercial technology, although there have been attempts to use it since at least 1890. In 2008, the first experimental wave farm was opened in Portugal, at the Aguçadoura Wave Park.

Physical Concepts

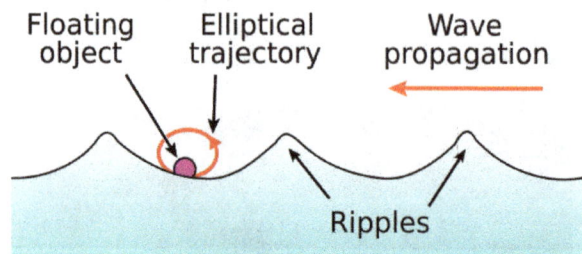

When an object bobs up and down on a ripple in a pond, it follows approximately an elliptical trajectory.

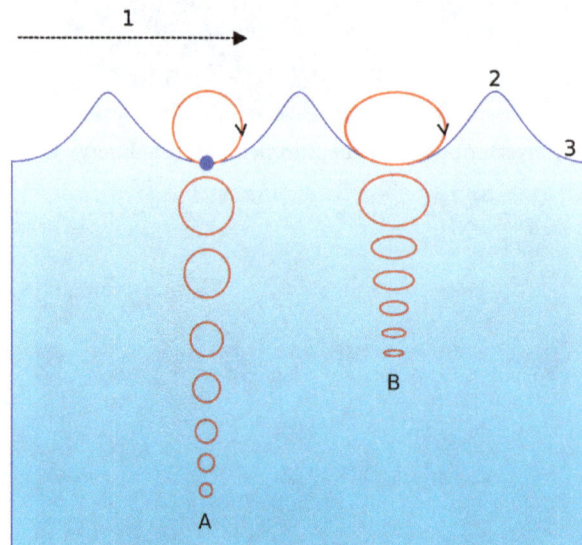

Motion of a particle in an ocean wave.

A = At deep water. The elliptical motion of fluid particles decreases rapidly with increasing depth below the surface.

B = At shallow water (ocean floor is now at B). The elliptical movement of a fluid particle flattens with decreasing depth.

1 = Propagation direction.

2 = Wave crest.

3 = Wave trough.

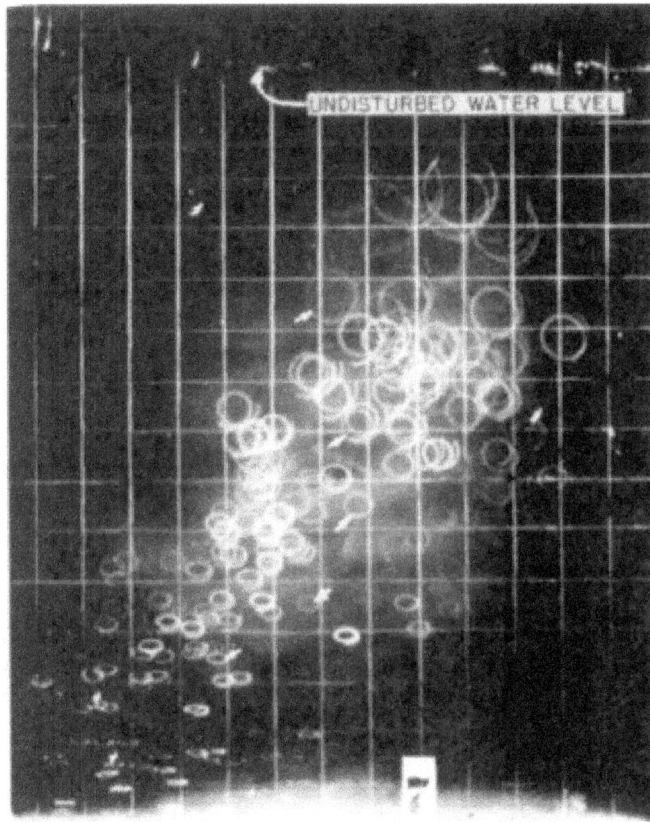

Photograph of the elliptical trajectories of water particles under a – progressive and periodic – surface gravity wave in a wave flume. The wave conditions are: mean water depth d = 2.50 ft (0.76 m), wave height H = 0.339 ft (0.103 m), wavelength λ = 6.42 ft (1.96 m), period T = 1.12 s.

Waves are generated by wind passing over the surface of the sea. As long as the waves propagate slower than the wind speed just above the waves, there is an energy transfer from the wind to the waves. Both air pressure differences between the upwind and the lee side of a wave crest, as well as friction on the water surface by the wind, making the water to go into the shear stress causes the growth of the waves.

Wave height is determined by wind speed, the duration of time the wind has been blowing, fetch (the distance over which the wind excites the waves) and by the depth and topography of the seafloor (which can focus or disperse the energy of the waves). A given wind speed has a matching practical limit over which time or distance will not produce larger waves. When this limit has been reached the sea is said to be "fully developed".

In general, larger waves are more powerful but wave power is also determined by wave speed, wavelength, and water density.

Oscillatory motion is highest at the surface and diminishes exponentially with depth. However, for standing waves (clapotis) near a reflecting coast, wave energy is also present as pressure oscillations at great depth, producing microseisms. These pressure fluctuations at greater depth are too small to be interesting from the point of view of wave power.

The waves propagate on the ocean surface, and the wave energy is also transported horizontally with the group velocity. The mean transport rate of the wave energy through a vertical plane of unit width, parallel to a wave crest, is called the wave energy flux (or wave power, which must not be confused with the actual power generated by a wave power device).

Wave Power Formula

In deep water where the water depth is larger than half the wavelength, the wave energy flux is

$$P = \frac{\rho g^2}{64\pi} H_{m0}^2 T_e \approx \left(0.5 \frac{\text{kW}}{\text{m}^3 \cdot \text{s}}\right) H_{m0}^2 T_e,$$

with P the wave energy flux per unit of wave-crest length, H_{m0} the significant wave height, T_e the wave energy period, ρ the water density and g the acceleration by gravity. The above formula states that wave power is proportional to the wave energy period and to the square of the wave height. When the significant wave height is given in metres, and the wave period in seconds, the result is the wave power in kilowatts (kW) per metre of wavefront length.

Example: Consider moderate ocean swells, in deep water, a few km off a coastline, with a wave height of 3 m and a wave energy period of 8 seconds. Using the formula to solve for power, we get

$$P \approx 0.5 \frac{\text{kW}}{\text{m}^3 \cdot \text{s}} (3 \cdot \text{m})^2 (8 \cdot \text{s}) \approx 36 \frac{\text{kW}}{\text{m}},$$

meaning there are 36 kilowatts of power potential per meter of wave crest.

In major storms, the largest waves offshore are about 15 meters high and have a period of about 15 seconds. According to the above formula, such waves carry about 1.7 MW of power across each metre of wavefront.

An effective wave power device captures as much as possible of the wave energy flux. As a result, the waves will be of lower height in the region behind the wave power device.

Wave Energy and Wave-energy Flux

In a sea state, the average(mean) energy density per unit area of gravity waves on the water surface is proportional to the wave height squared, according to linear wave theory:

$$E = \frac{1}{8}\rho g H_{m0}^2,$$

where E is the mean wave energy density per unit horizontal area (J/m²), the sum of kinetic and potential energy density per unit horizontal area. The potential energy density is equal to the kinetic energy, both contributing half to the wave energy density E, as can be expected from the equipartition theorem. In ocean waves, surface tension effects are negligible for wavelengths above a few decimetres.

As the waves propagate, their energy is transported. The energy transport velocity is the group velocity. As a result, the wave energy flux, through a vertical plane of unit width perpendicular to the wave propagation direction, is equal to:

$$P = Ec_g,$$

with c_g the group velocity (m/s). Due to the dispersion relation for water waves under the action of gravity, the group velocity depends on the wavelength λ, or equivalently, on the wave period T. Further, the dispersion relation is a function of the water depth h. As a result, the group velocity behaves differently in the limits of deep and shallow water, and at intermediate depths:

Properties of gravity waves on the surface of deep water, shallow water and at intermediate depth, according to linear wave theory					
quantity	symbol	units	deep water ($h > \tfrac{1}{2}\lambda$)	shallow water ($h < 0.05\,\lambda$)	intermediate depth (all λ and h)
phase velocity	$c_p = \dfrac{\lambda}{T} = \dfrac{\omega}{k}$	m / s	$\dfrac{g}{2\pi}T$	$\sqrt{gh}$	$\sqrt{\dfrac{g\lambda}{2\pi}\tanh\left(\dfrac{2\pi h}{\lambda}\right)}$
group velocity	$c_g = c_p^2\,\dfrac{\partial(\lambda/c_p)}{\partial\lambda} = \dfrac{\partial\omega}{\partial k}$	m / s	$\dfrac{g}{4\pi}T$	$\sqrt{gh}$	$\dfrac{1}{2}c_p\left(1+\dfrac{4\pi h}{\lambda}\dfrac{1}{\sinh\left(\dfrac{4\pi h}{\lambda}\right)}\right)$
ratio	$\dfrac{c_g}{c_p}$	–	$\dfrac{1}{2}$	1	$\dfrac{1}{2}\left(1+\dfrac{4\pi h}{\lambda}\dfrac{1}{\sinh\left(\dfrac{4\pi h}{\lambda}\right)}\right)$
wavelength	λ	m	$\dfrac{g}{2\pi}T^2$	$T\sqrt{gh}$	for given period T, the solution of: $\left(\dfrac{2\pi}{T}\right)^2 = \dfrac{2\pi g}{\lambda}\tanh\left(\dfrac{2\pi h}{\lambda}\right)$
general					
wave energy density	E	J / m²	$\dfrac{1}{16}\rho g H_{m0}^2$		

wave energy flux	P	W / m	$E\,c_g$
angular frequency	ω	rad / s	$\dfrac{2\pi}{T}$
wavenumber	k	rad / m	$\dfrac{2\pi}{\lambda}$

Deep-water Characteristics and Opportunities

Deep water corresponds with a water depth larger than half the wavelength, which is the common situation in the sea and ocean. In deep water, longer-period waves propagate faster and transport their energy faster. The deep-water group velocity is half the phase velocity. In shallow water, for wavelengths larger than about twenty times the water depth, as found quite often near the coast, the group velocity is equal to the phase velocity.

History

The first known patent to use energy from ocean waves dates back to 1799, and was filed in Paris by Girard and his son. An early application of wave power was a device constructed around 1910 by Bochaux-Praceique to light and power his house at Royan, near Bordeaux in France. It appears that this was the first oscillating water-column type of wave-energy device. From 1855 to 1973 there were already 340 patents filed in the UK alone.

Modern scientific pursuit of wave energy was pioneered by Yoshio Masuda's experiments in the 1940s. He has tested various concepts of wave-energy devices at sea, with several hundred units used to power navigation lights. Among these was the concept of extracting power from the angular motion at the joints of an articulated raft, which was proposed in the 1950s by Masuda.

A renewed interest in wave energy was motivated by the oil crisis in 1973. A number of university researchers re-examined the potential to generate energy from ocean waves, among whom notably were Stephen Salter from the University of Edinburgh, Kjell Budal and Johannes Falnes from Norwegian Institute of Technology (now merged into Norwegian University of Science and Technology), Michael E. McCormick from U.S. Naval Academy, David Evans from Bristol University, Michael French from University of Lancaster, Nick Newman and C. C. Mei from MIT.

Stephen Salter's 1974 invention became known as Salter's duck or *nodding duck*, although it was officially referred to as the Edinburgh Duck. In small scale controlled tests, the Duck's curved cam-like body can stop 90% of wave motion and can convert 90% of that to electricity giving 81% efficiency.

In the 1980s, as the oil price went down, wave-energy funding was drastically reduced. Nevertheless, a few first-generation prototypes were tested at sea. More recently, following the issue of climate change, there is again a growing interest worldwide for renewable energy, including wave energy.

The world's first marine energy test facility was established in 2003 to kick start the development of a wave and tidal energy industry in the UK. Based in Orkney, Scotland, the European Marine

Energy Centre (EMEC) has supported the deployment of more wave and tidal energy devices than at any other single site in the world. EMEC provides a variety of test sites in real sea conditions. It's grid connected wave test site is situated at Billia Croo, on the western edge of the Orkney mainland, and is subject to the full force of the Atlantic Ocean with seas as high as 19 metres recorded at the site. Wave energy developers currently testing at the centre include Aquamarine Power, Pelamis Wave Power, ScottishPower Renewables and Wello.

Modern technology

Wave power devices are generally categorized by the method used to capture the energy of the waves, by location and by the power take-off system. Locations are shoreline, nearshore and offshore. Types of power take-off include: hydraulic ram, elastomeric hose pump, pump-to-shore, hydroelectric turbine, air turbine, and linear electrical generator. When evaluating wave energy as a technology type, it is important to distinguish between the four most common approaches: point absorber buoys, surface attenuators, oscillating water columns, and overtopping devices.

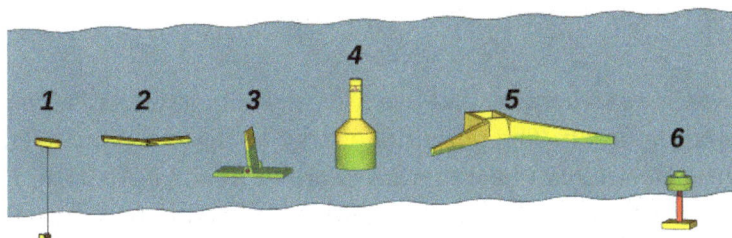

Generic wave energy concepts: 1. Point absorber, 2. Attenuator, 3. Oscillating wave surge converter, 4. Oscillating water column, 5. Overtopping device, 6. Submerged pressure differential

Point Absorber Buoy

This device floats on the surface of the water, held in place by cables connected to the seabed. Buoys use the rise and fall of swells to drive hydraulic pumps and generate electricity. EMF generated by electrical transmission cables and acoustics of these devices may be a concern for marine organisms. The presence of the buoys may affect fish, marine mammals, and birds as potential minor collision risk and roosting sites. Potential also exists for entanglement in mooring lines. Energy removed from the waves may also affect the shoreline, resulting in a recommendation that sites remain a considerable distance from the shore.

Surface Attenuator

These devices act similarly to point absorber buoys, with multiple floating segments connected to one another and are oriented perpendicular to incoming waves. A flexing motion is created by swells that drive hydraulic pumps to generate electricity. Environmental effects are similar to those of point absorber buoys, with an additional concern that organisms could be pinched in the joints.

Oscillating Wave Surge Converter

These devices typically have one end fixed to a structure or the seabed while the other end is free to move. Energy is collected from the relative motion of the body compared to the fixed point.

Oscillating wave surge converters often come in the form of floats, flaps, or membranes. Environmental concerns include minor risk of collision, artificial reefing near the fixed point, EMF effects from subsea cables, and energy removal effecting sediment transport. Some of these designs incorporate parabolic reflectors as a means of increasing the wave energy at the point of capture. These capture systems use the rise and fall motion of waves to capture energy. Once the wave energy is captured at a wave source, power must be carried to the point of use or to a connection to the electrical grid by transmission power cables.

Oscillating Water Column

Oscillating Water Column devices can be located on shore or in deeper waters offshore. With an air chamber integrated into the device, swells compress air in the chambers forcing air through an air turbine to create electricity. Significant noise is produced as air is pushed through the turbines, potentially affecting birds and other marine organisms within the vicinity of the device. There is also concern about marine organisms getting trapped or entangled within the air chambers.

Overtopping Device

Overtopping devices are long structures that use wave velocity to fill a reservoir to a greater water level than the surrounding ocean. The potential energy in the reservoir height is then captured with low-head turbines. Devices can be either on shore or floating offshore. Floating devices will have environmental concerns about the mooring system affecting benthic organisms, organisms becoming entangled, or EMF effects produced from subsea cables. There is also some concern regarding low levels of turbine noise and wave energy removal affecting the nearfield habitat.

Environmental Effects

Common environmental concerns associated with marine energy developments include:

- The risk of marine mammals and fish being struck by tidal turbine blades;

- The effects of EMF and underwater noise emitted from operating marine energy devices;

- The physical presence of marine energy projects and their potential to alter the behavior of marine mammals, fish, and seabirds with attraction or avoidance;

- The potential effect on nearfield and farfield marine environment and processes such as sediment transport and water quality.

The Tethys database provides access to scientific literature and general information on the potential environmental effects of wave energy.

Potential

The worldwide resource of wave energy has been estimated to be greater than 2 TW. Locations with the most potential for wave power include the western seaboard of Europe, the northern coast of the UK, and the Pacific coastlines of North and South America, Southern Africa, Australia, and New Zealand. The north and south temperate zones have the best sites for capturing wave power. The prevailing westerlies in these zones blow strongest in winter.

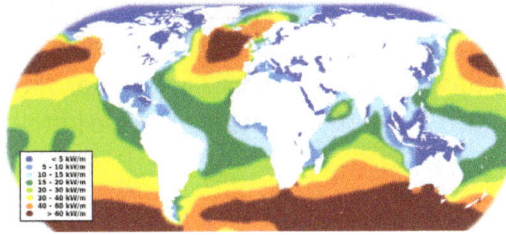

World wave energy resource map

Challenges

There is a potential impact on the marine environment. Noise pollution, for example, could have negative impact if not monitored, although the noise and visible impact of each design vary greatly. Other biophysical impacts (flora and fauna, sediment regimes and water column structure and flows) of scaling up the technology is being studied. In terms of socio-economic challenges, wave farms can result in the displacement of commercial and recreational fishermen from productive fishing grounds, can change the pattern of beach sand nourishment, and may represent hazards to safe navigation. Waves generate about 2,700 gigawatts of power. Of those 2,700 gigawatts, only about 500 gigawatts can be captured with the current technology.

Wave Farms

Portugal

- The Aguçadoura Wave Farm was the world's first wave farm. It was located 5 km (3 mi) offshore near Póvoa de Varzim, north of Porto, Portugal. The farm was designed to use three Pelamis wave energy converters to convert the motion of the ocean surface waves into electricity, totalling to 2.25 MW in total installed capacity. The farm first generated electricity in July 2008 and was officially opened on September 23, 2008, by the Portuguese Minister of Economy. The wave farm was shut down two months after the official opening in November 2008 as a result of the financial collapse of Babcock & Brown due to the global economic crisis. The machines were off-site at this time due to technical problems, and although resolved have not returned to site and were subsequently scrapped in 2011 as the technology had moved on to the P2 variant as supplied to Eon and Scottish Power Renewables. A second phase of the project planned to increase the installed capacity to 21 MW using a further 25 Pelamis machines is in doubt following Babcock's financial collapse.

United Kingdom

- Funding for a 3 MW wave farm in Scotland was announced on February 20, 2007, by the Scottish Executive, at a cost of over 4 million pounds, as part of a £13 million funding package for marine power in Scotland. The first machine was launched in May 2010.

- A facility known as Wave hub has been constructed off the north coast of Cornwall, England, to facilitate wave energy development. The Wave hub will act as giant extension cable, allowing arrays of wave energy generating devices to be connected to the electricity grid. The Wave hub will initially allow 20 MW of capacity to be connected, with potential expansion to 40 MW. Four device manufacturers have so far expressed interest in connecting to the Wave hub. The scientists have calculated that wave energy gathered at Wave Hub will be enough to power up to 7,500 households. The site has the potential to save greenhouse gas emissions of about 300,000 tons of carbon dioxide in the next 25 years.

Australia

- A CETO wave farm off the coast of Western Australia has been operating to prove commercial viability and, after preliminary environmental approval, underwent further development. In early 2015 a $100 million, multi megawatt system was connected to the grid, with all the electricity being bought to power HMAS Stirling naval base. Two fully submerged buoys which are anchored to the seabed, transmit the energy from the ocean swell through hydraulic pressure onshore; to drive a generator for electricity, and also to produce fresh water. As of 2015[update] a third buoy is planned for installation.

- Ocean Power Technologies (OPT Australasia Pty Ltd) is developing a wave farm connected to the grid near Portland, Victoria through a 19 MW wave power station. The project has received an AU $66.46 million grant from the Federal Government of Australia.

- Oceanlinx will deploy a commercial scale demonstrator off the coast of South Australia at Port MacDonnell before the end of 2013. This device, the *greenWAVE*, has a rated electrical capacity of 1MW. This project has been supported by ARENA through the Emerging Renewables Program. The *greenWAVE* device is a bottom standing gravity structure, that does not require anchoring or seabed preparation and with no moving parts below the surface of the water.

United States

- Reedsport, Oregon – a commercial wave park on the west coast of the United States located 2.5 miles offshore near Reedsport, Oregon. The first phase of this project is for ten PB150 PowerBuoys, or 1.5 megawatts. The Reedsport wave farm was scheduled for installation spring 2013. In 2013, the project has ground to a halt because of legal and technical problems.

- Kaneohe Bay Oahu, Hawai - Navy's Wave Energy Test Site (WETS) currently testing the Azura wave power device

Patents

- U.S. Patent 8,806,865 — 2011 *Ocean wave energy harnessing device* – Pelamis/Salter's Duck Hybrid patent

- U.S. Patent 3,928,967 — 1974 *Apparatus and method of extracting wave energy* – The original "Salter's Duck" patent

- U.S. Patent 4,134,023 — 1977 *Apparatus for use in the extraction of energy from waves on water* – Salter's method for improving "duck" efficiency

- U.S. Patent 6,194,815 — 1999 *Piezoelectric rotary electrical energy generator*

- Wave energy converters utilizing pressure differences US 20040217597 A1 — 2004 *Wave energy converters utilizing pressure differences*

References

- Taylor, John W R (1974). Jane's All the World's Aircraft 1974-75. London: Jane's Yearbooks. p. 573. ISBN 0 354 00502 2.

- Spellman, Frank R. (2013). Safe Work Practices for Green Energy Jobs (first ed.). DEStech Publications. p. 323. ISBN 978-1-60595-075-4. Retrieved 29 December 2014.

- Solar Cells and their Applications Second Edition, Lewis Fraas, Larry Partain, Wiley, 2010, ISBN 978-0-470-44633-1 , Section10.2.

- Collings AF, Critchley C. Artificial Photosynthesis. From Basic Biology to Industrial Application. Wiley-VCH. Weinheim (2005) p. x ISBN 3-527-31090-8 doi:10.1002/3527606742.

- History of Wind Energy in Cutler J. Cleveland,(ed) Encyclopedia of Energy Vol.6, Elsevier, ISBN 978-1-60119-433-6, 2007, pp. 421–422

- Tucker, M.J.; Pitt, E.G. (2001). "2". In Bhattacharyya, R.; McCormick, M.E. Waves in ocean engineering (1st ed.). Oxford: Elsevier. pp. 35–36. ISBN 0080435661.

- Holthuijsen, Leo H. (2007). Waves in oceanic and coastal waters. Cambridge: Cambridge University Press. ISBN 0-521-86028-8.

- R. G. Dean & R. A. Dalrymple (1991). Water wave mechanics for engineers and scientists. Advanced Series on Ocean Engineering. 2. World Scientific, Singapore. ISBN 978-981-02-0420-4.

- Worland, Justin (4 April 2016). "After years of torrid growth, residential solar power faces serious growing pains". Time. Vol. 187 no. 12. p. 24. Retrieved 10 April 2016 – via Issuu.

- "Electric cars not solar panels, says Environment Commissioner". Parliamentary Commissioner for the Environment. 22 March 2016. Retrieved 23 March 2016.

- Started in August 2001, the Jaisalmer based facility crossed 1,000 MW capacity to achieve this milestone. Business-standard.com (11 May 2012). Retrieved on 20 July 2016.

- Inadequate transmission lines keeping some Maine wind power off the grid – The Portland Press Herald / Maine Sunday Telegram. Pressherald.com (4 August 2013). Retrieved on 2016-07-20.

- $7.3 billion in public health savings seen in 2015 from wind energy cutting air pollution. Awea.org (29 March 2016). Retrieved on 2016-07-20.

Significant Innovations of Green Energy

Constant research and technical development has led to the adoption of green energy compatible technology in machinery like cars and buses as well as in the design of buildings and recreational centres. The chapter strategically encompasses and incorporates the major components and key concepts of green energy, providing a complete understanding.

Geothermal Energy

Geothermal energy is thermal energy generated and stored in the Earth. Thermal energy is the energy that determines the temperature of matter. The geothermal energy of the Earth's crust originates from the original formation of the planet and from radioactive decay of materials (in currently uncertain but possibly roughly equal proportions). The geothermal gradient, which is the difference in temperature between the core of the planet and its surface, drives a continuous conduction of thermal energy in the form of heat from the core to the surface.

Steam rising from the Nesjavellir Geothermal Power Station in Iceland.

Earth's internal heat is thermal energy generated from radioactive decay and continual heat loss from Earth's formation. Temperatures at the core–mantle boundary may reach over 4000 °C (7,200 °F). The high temperature and pressure in Earth's interior cause some rock to melt and solid mantle to behave plastically, resulting in portions of mantle convecting upward since it is lighter than the surrounding rock. Rock and water is heated in the crust, sometimes up to 370 °C (700 °F).

From hot springs, geothermal energy has been used for bathing since Paleolithic times and for space heating since ancient Roman times, but it is now better known for electricity generation. Worldwide, 11,700 megawatts (MW) of geothermal power is online in 2013. An additional 28 gigawatts of direct geothermal heating capacity is installed for district heating, space heating, spas, industrial processes, desalination and agricultural applications in 2010.

Geothermal power is cost-effective, reliable, sustainable, and environmentally friendly, but has historically been limited to areas near tectonic plate boundaries. Recent technological advances have dramatically expanded the range and size of viable resources, especially for applications such as home heating, opening a potential for widespread exploitation. Geothermal wells release greenhouse gases trapped deep within the earth, but these emissions are much lower per energy unit than those of fossil fuels. As a result, geothermal power has the potential to help mitigate global warming if widely deployed in place of fossil fuels.

The Earth's geothermal resources are theoretically more than adequate to supply humanity's energy needs, but only a very small fraction may be profitably exploited. Drilling and exploration for deep resources is very expensive. Forecasts for the future of geothermal power depend on assumptions about technology, energy prices, subsidies, and interest rates. Pilot programs like EWEB's customer opt in Green Power Program show that customers would be willing to pay a little more for a renewable energy source like geothermal. But as a result of government assisted research and industry experience, the cost of generating geothermal power has decreased by 25% over the past two decades. In 2001, geothermal energy costs between two and ten US cents per kWh.

History

The oldest known pool fed by a hot spring, built in the Qin dynasty in the 3rd century BCE.

Hot springs have been used for bathing at least since Paleolithic times. The oldest known spa is a stone pool on China's Lisan mountain built in the Qin Dynasty in the 3rd century BC, at the same site where the Huaqing Chi palace was later built. In the first century AD, Romans conquered *Aquae Sulis*, now Bath, Somerset, England, and used the hot springs there to feed public baths and underfloor heating. The admission fees for these baths probably represent the first commercial use of geothermal power. The world's oldest geothermal district heating system in Chaudes-Aigues,

France, has been operating since the 14th century. The earliest industrial exploitation began in 1827 with the use of geyser steam to extract boric acid from volcanic mud in Larderello, Italy.

In 1892, America's first district heating system in Boise, Idaho was powered directly by geothermal energy, and was copied in Klamath Falls, Oregon in 1900. A deep geothermal well was used to heat greenhouses in Boise in 1926, and geysers were used to heat greenhouses in Iceland and Tuscany at about the same time. Charlie Lieb developed the first downhole heat exchanger in 1930 to heat his house. Steam and hot water from geysers began heating homes in Iceland starting in 1943.

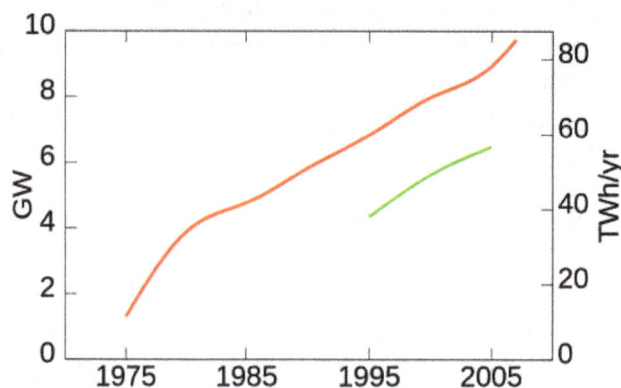

Global geothermal electric capacity. Upper red line is installed capacity; lower green line is realized production.

In the 20th century, demand for electricity led to the consideration of geothermal power as a generating source. Prince Piero Ginori Conti tested the first geothermal power generator on 4 July 1904, at the same Larderello dry steam field where geothermal acid extraction began. It successfully lit four light bulbs. Later, in 1911, the world's first commercial geothermal power plant was built there. It was the world's only industrial producer of geothermal electricity until New Zealand built a plant in 1958. In 2012, it produced some 594 megawatts.

Lord Kelvin invented the heat pump in 1852, and Heinrich Zoelly had patented the idea of using it to draw heat from the ground in 1912. But it was not until the late 1940s that the geothermal heat pump was successfully implemented. The earliest one was probably Robert C. Webber's home-made 2.2 kW direct-exchange system, but sources disagree as to the exact timeline of his invention. J. Donald Kroeker designed the first commercial geothermal heat pump to heat the Commonwealth Building (Portland, Oregon) and demonstrated it in 1946. Professor Carl Nielsen of Ohio State University built the first residential open loop version in his home in 1948. The technology became popular in Sweden as a result of the 1973 oil crisis, and has been growing slowly in worldwide acceptance since then. The 1979 development of polybutylene pipe greatly augmented the heat pump's economic viability.

In 1960, Pacific Gas and Electric began operation of the first successful geothermal electric power plant in the United States at The Geysers in California. The original turbine lasted for more than 30 years and produced 11 MW net power.

The binary cycle power plant was first demonstrated in 1967 in the USSR and later introduced to the US in 1981. This technology allows the generation of electricity from much lower temperature resources than previously. In 2006, a binary cycle plant in Chena Hot Springs, Alaska, came online, producing electricity from a record low fluid temperature of 57 °C (135 °F).

Direct Usage

Direct Use Data 2015	
Country	**Usage (MWt) 2015**
United States	17,415.91
Philippines	3.30
Indonesia	2.30
Mexico	155.82
Italy	1,014.00
New Zealand	487.45
Iceland	2,040.00
Japan	2,186.17
Iran	81.50
El Salvador	3.36
Kenya	22.40
Costa Rica	1.00
Russia	308.20
Turkey	2,886.30
Papua-New Guinea	0.10
Guatemala	2.31
Portugal	35.20
China	17,870.00
France	2,346.90
Ethiopia	2.20
Germany	2,848.60
Austria	903.40
Australia	16.09
Thailand	128.51

Electricity

The International Geothermal Association (IGA) has reported that 10,715 megawatts (MW) of geothermal power in 24 countries is online, which was expected to generate 67,246 GWh of electricity in 2010. This represents a 20% increase in online capacity since 2005. IGA projects growth to 18,500 MW by 2015, due to the projects presently under consideration, often in areas previously assumed to have little exploitable resource.

In 2010, the United States led the world in geothermal electricity production with 3,086 MW of installed capacity from 77 power plants. The largest group of geothermal power plants in the world is located at The Geysers, a geothermal field in California. The Philippines is the second highest producer, with 1,904 MW of capacity online. Geothermal power makes up approximately 27% of Philippine electricity generation.

Installed geothermal electric capacity				
Country	Capacity (MW) 2007	Capacity (MW) 2010	Percentage of national electricity production	Percentage of global geothermal production
United States	2687	3086	0.3	29
Philippines	1969.7	1904	27	18
Indonesia	992	1197	3.7	11
Mexico	953	958	3	9
Italy	810.5	843	1.5	8
New Zealand	471.6	628	10	6
Iceland	421.2	575	30	5
Japan	535.2	536	0.1	5
Iran	250	250		
El Salvador	204.2	204	25	
Kenya	128.8	167	11.2	
Costa Rica	162.5	166	14	
Nicaragua	87.4	88	10	
Russia	79	82		
Turkey	38	82		
Papua-New Guinea	56	56		
Guatemala	53	52		
Portugal	23	29		
China	27.8	24		
France	14.7	16		
Ethiopia	7.3	7.3		
Germany	8.4	6.6		
Austria	1.1	1.4		
Australia	0.2	1.1		
Thailand	0.3	0.3		
TOTAL	**9,981.9**	**10,959.7**		

Geothermal electric plants were traditionally built exclusively on the edges of tectonic plates where high temperature geothermal resources are available near the surface. The development of binary cycle power plants and improvements in drilling and extraction technology enable enhanced geothermal systems over a much greater geographical range. Demonstration projects are operational in Landau-Pfalz, Germany, and Soultz-sous-Forêts, France, while an earlier effort in Basel, Switzerland was shut down after it triggered earthquakes. Other demonstration projects are under construction in Australia, the United Kingdom, and the United States of America.

The thermal efficiency of geothermal electric plants is low, around 10–23%, because geothermal fluids do not reach the high temperatures of steam from boilers. The laws of thermodynamics limits the efficiency of heat engines in extracting useful energy. Exhaust heat is wasted, unless it can be used directly and locally, for example in greenhouses, timber mills, and district heating. System efficiency does not materially affect operational costs as it would for plants that use fuel, but it does affect return on the capital used to build the plant. In order to produce more energy than the pumps consume, electricity generation requires relatively hot fields and specialized heat cycles. Because geothermal power does not rely on variable sources of energy, unlike, for example, wind or solar, its capacity factor can be quite large – up to 96% has been demonstrated. The global average was 73% in 2005.

Types

Geothermal energy comes in either *vapor-dominated* or *liquid-dominated* forms. Larderello and The Geysers are vapor-dominated. Vapor-dominated sites offer temperatures from 240 to 300 °C that produce superheated steam.

Liquid-dominated Plants

Liquid-dominated reservoirs (LDRs) were more common with temperatures greater than 200 °C (392 °F) and are found near young volcanoes surrounding the Pacific Ocean and in rift zones and hot spots. *Flash plants* are the common way to generate electricity from these sources. Pumps are generally not required, powered instead when the water turns to steam. Most wells generate 2-10MWe. Steam is separated from liquid via cyclone separators, while the liquid is returned to the reservoir for reheating/reuse. As of 2013, the largest liquid system is Cerro Prieto in Mexico, which generates 750 MWe from temperatures reaching 350 °C (662 °F). The Salton Sea field in Southern California offers the potential of generating 2000 MWe.

Lower temperature LDRs (120–200 °C) require pumping. They are common in extensional terrains, where heating takes place via deep circulation along faults, such as in the Western US and Turkey. Water passes through a heat exchanger in a Rankine cycle binary plant. The water vaporizes an organic working fluid that drives a turbine. These binary plants originated in the Soviet Union in the late 1960s and predominate in new US plants. Binary plants have no emissions.

Thermal Energy

Lower temperature sources produce the energy equivalent of 100M BBL per year. Sources with temperatures of 30–150 °C are used without conversion to electricity as district heating, greenhouses, fisheries, mineral recovery, industrial process heating and bathing in 75 countries. Heat pumps extract energy from shallow sources at 10–20 °C in 43 countries for use in space heating and cooling. Home heating is the fastest-growing means of exploiting geothermal energy, with global annual growth rate of 30% in 2005 and 20% in 2012.

Approximately 270 petajoules (PJ) of geothermal heating was used in 2004. More than half went for space heating, and another third for heated pools. The remainder supported industrial and agricultural applications. Global installed capacity was 28 GW, but capacity factors tend to be low (30% on average) since heat is mostly needed in winter. Some 88 PJ for space heating was extracted by an estimated 1.3 million geothermal heat pumps with a total capacity of 15 GW.

Heat for these purposes may also be extracted from co-generation at a geothermal electrical plant.

Heating is cost-effective at many more sites than electricity generation. At natural hot springs or geysers, water can be piped directly into radiators. In hot, dry ground, earth tubes or downhole heat exchangers can collect the heat. However, even in areas where the ground is colder than room temperature, heat can often be extracted with a geothermal heat pump more cost-effectively and cleanly than by conventional furnaces. These devices draw on much shallower and colder resources than traditional geothermal techniques. They frequently combine functions, including air conditioning, seasonal thermal energy storage, solar energy collection, and electric heating. Heat pumps can be used for space heating essentially anywhere.

Iceland is the world leader in direct applications. Some 92.5% of its homes are heated with geothermal energy, saving Iceland over $100 million annually in avoided oil imports. Reykjavík, Iceland has the world's biggest district heating system. Once known as the most polluted city in the world, it is now one of the cleanest.

Enhanced Geothermal

Enhanced geothermal systems (EGS) actively inject water into wells to be heated and pumped back out. The water is injected under high pressure to expand existing rock fissures to enable the water to freely flow in and out. The technique was adapted from oil and gas extraction techniques. However, the geologic formations are deeper and no toxic chemicals are used, reducing the possibility of environmental damage. Drillers can employ directional drilling to expand the size of the reservoir.

Small-scale EGS have been installed in the Rhine Graben at Soultz-sous-Forêts in France and at Landau and Insheim in Germany.

Economics

Geothermal power requires no fuel (except for pumps), and is therefore immune to fuel cost fluctuations. However, capital costs are significant. Drilling accounts for over half the costs, and exploration of deep resources entails significant risks. A typical well doublet (extraction and injection wells) in Nevada can support 4.5 megawatts (MW) and costs about $10 million to drill, with a 20% failure rate.

A power plant at The Geysers

In total, electrical plant construction and well drilling cost about €2–5 million per MW of electrical capacity, while the break–even price is 0.04–0.10 € per kW·h. Enhanced geothermal systems tend to be on the high side of these ranges, with capital costs above $4 million per MW and break–even above $0.054 per kW·h in 2007. Direct heating applications can use much shallower wells with lower temperatures, so smaller systems with lower costs and risks are feasible. Residential geothermal heat pumps with a capacity of 10 kilowatt (kW) are routinely installed for around $1–3,000 per kilowatt. District heating systems may benefit from economies of scale if demand is geographically dense, as in cities and greenhouses, but otherwise piping installation dominates capital costs. The capital cost of one such district heating system in Bavaria was estimated at somewhat over 1 million € per MW. Direct systems of any size are much simpler than electric generators and have lower maintenance costs per kW·h, but they must consume electricity to run pumps and compressors. Some governments subsidize geothermal projects.

Geothermal power is highly scalable: from a rural village to an entire city.

The most developed geothermal field in the United States is The Geysers in Northern California.

Geothermal projects have several stages of development. Each phase has associated risks. At the early stages of reconnaissance and geophysical surveys, many projects are cancelled, making that phase unsuitable for traditional lending. Projects moving forward from the identification, exploration and exploratory drilling often trade equity for financing.

Resources

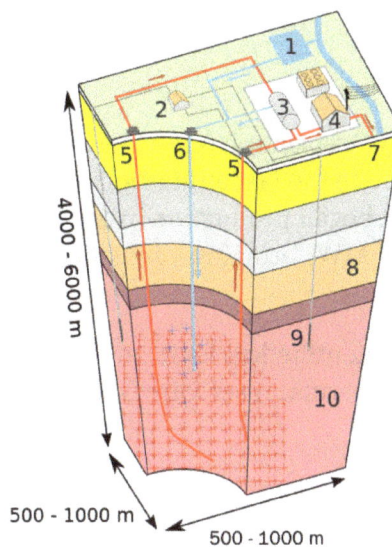

Enhanced geothermal system 1:Reservoir 2:Pump house 3:Heat exchanger 4:Turbine hall 5:Pro-duction well 6:Injection well 7:Hot water to district heating 8:Porous sediments 9:Observation well 10:Crystalline bedrock

The Earth's internal thermal energy flows to the surface by conduction at a rate of 44.2 terawatts (TW), and is replenished by radioactive decay of minerals at a rate of 30 TW. These power rates are more than double humanity's current energy consumption from all primary sources, but most of this energy flow is not recoverable. In addition to the internal heat flows, the top layer of the surface to a depth of 10 meters (33 ft) is heated by solar energy during the summer, and releases that energy and cools during the winter.

Outside of the seasonal variations, the geothermal gradient of temperatures through the crust is 25–30 °C (77–86 °F) per kilometer of depth in most of the world. The conductive heat flux averages 0.1 MW/km². These values are much higher near tectonic plate boundaries where the crust is thinner. They may be further augmented by fluid circulation, either through magma conduits, hot springs, hydrothermal circulation or a combination of these.

A geothermal heat pump can extract enough heat from shallow ground anywhere in the world to provide home heating, but industrial applications need the higher temperatures of deep resources. The thermal efficiency and profitability of electricity generation is particularly sensitive to temperature. The most demanding applications receive the greatest benefit from a high natural heat flux, ideally from using a hot spring. The next best option is to drill a well into a hot aquifer. If no adequate aquifer is available, an artificial one may be built by injecting water to hydraulically fracture the bedrock. This last approach is called hot dry rock geothermal energy in Europe, or enhanced geothermal systems in North America. Much greater potential may be available from this approach than from conventional tapping of natural aquifers.

Estimates of the potential for electricity generation from geothermal energy vary sixfold, from .035to2TW depending on the scale of investments. Upper estimates of geothermal resources assume enhanced geothermal wells as deep as 10 kilometres (6 mi), whereas existing geothermal wells are rarely more than 3 kilometres (2 mi) deep. Wells of this depth are now common in the petroleum industry. The deepest research well in the world, the Kola superdeep borehole, is 12 kilometres (7 mi) deep.

Production

According to the Geothermal Energy Association (GEA) installed geothermal capacity in the United States grew by 5%, or 147.05 MW, ce the last annual survey in March 2012. This increase came from seven geothermal projects that began production in 2012. GEA also revised its 2011 estimate of installed capacity upward by 128 MW, bringing current installed U.S. geothermal capacity to 3,386 MW.

Myanmar Engineering Society has identified at least 39 locations capable of geothermal power production and some of these hydrothermal reservoirs lie quite close to Yangon which is a significant underutilized resource.

Renewability and Sustainability

Geothermal power is considered to be renewable because any projected heat extraction is small compared to the Earth's heat content. The Earth has an internal heat content of 10^{31} joules ($3 \cdot 10^{15}$ TW·hr), approximately 100 billion times current (2010) worldwide annual energy consumption. About 20% of this is residual heat from planetary accretion, and the remainder is attributed to higher radioactive decay rates that existed in the past. Natural heat flows are not in equilibrium, and the planet is slowly cooling down on geologic timescales. Human extraction taps a minute fraction of the natural outflow, often without accelerating it.

Geothermal power is also considered to be sustainable thanks to its power to sustain the Earth's intricate ecosystems. By using geothermal sources of energy present generations of humans will not endanger the capability of future generations to use their own resources to the same amount that those energy sources are presently used. Further, due to its low emissions geothermal energy is considered to have excellent potential for mitigation of global warming.

Even though geothermal power is globally sustainable, extraction must still be monitored to avoid local depletion. Over the course of decades, individual wells draw down local temperatures and water levels until a new equilibrium is reached with natural flows. The three oldest sites, at Larderello, Wairakei, and the Geysers have experienced reduced output because of local depletion. Heat and water, in uncertain proportions, were extracted faster than they were replenished. If production is reduced and water is reinjected, these wells could theoretically recover their full potential. Such mitigation strategies have already been implemented at some sites. The long-term sustainability of geothermal energy has been demonstrated at the Lardarello field in Italy since 1913, at the Wairakei field in New Zealand since 1958, and at The Geysers field in California since 1960.

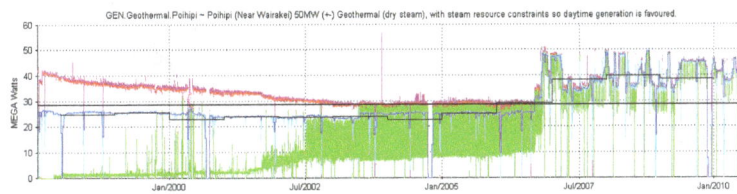

Electricity Generation at Poihipi, New Zealand.

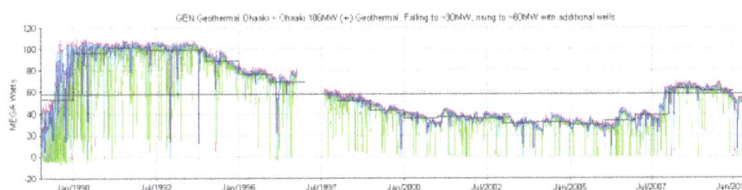

Electricity Generation at Ohaaki, New Zealand.

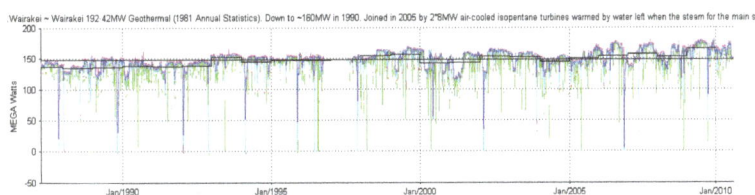

Electricity Generation at Wairakei, New Zealand.

Falling electricity production may be boosted through drilling additional supply boreholes, as at Poihipi and Ohaaki. The Wairakei power station has been running much longer, with its first unit commissioned in November 1958, and it attained its peak generation of 173MW in 1965, but already the supply of high-pressure steam was faltering, in 1982 being derated to intermediate pressure and the station managing 157MW. Around the start of the 21st century it was managing about 150MW, then in 2005 two 8MW isopentane systems were added, boosting the station's output by about 14MW. Detailed data are unavailable, being lost due to re-organisations. One such re-organisation in 1996 causes the absence of early data for Poihipi (started 1996), and the gap in 1996/7 for Wairakei and Ohaaki; half-hourly data for Ohaaki's first few months of operation are also missing, as well as for most of Wairakei's history.

Environmental Effects

Geothermal power station in the Philippines

Fluids drawn from the deep earth carry a mixture of gases, notably carbon dioxide (CO_2), hydrogen sulfide (H_2S), methane (CH_4) and ammonia (NH_3). These pollutants contribute to global warming, acid rain, and noxious smells if released. Existing geothermal electric plants emit an average of 122 kilograms (269 lb) of CO_2 per megawatt-hour (MW·h) of electricity, a small fraction of the emission intensity of conventional fossil fuel plants. Plants that experience high levels of acids and volatile chemicals are usually equipped with emission-control systems to reduce the exhaust.

Krafla Geothermal Station in northeast Iceland

In addition to dissolved gases, hot water from geothermal sources may hold in solution trace amounts of toxic elements such as mercury, arsenic, boron, and antimony. These chemicals precipitate as the water cools, and can cause environmental damage if released. The modern practice of injecting cooled geothermal fluids back into the Earth to stimulate production has the side benefit of reducing this environmental risk.

Direct geothermal heating systems contain pumps and compressors, which may consume energy from a polluting source. This parasitic load is normally a fraction of the heat output, so it is always less polluting than electric heating. However, if the electricity is produced by burning fossil fuels, then the net emissions of geothermal heating may be comparable to directly burning the fuel for heat. For example, a geothermal heat pump powered by electricity from a combined cycle natural gas plant would produce about as much pollution as a natural gas condensing furnace of the same size. Therefore, the environmental value of direct geothermal heating applications is highly dependent on the emissions intensity of the neighboring electric grid.

Plant construction can adversely affect land stability. Subsidence has occurred in the Wairakei field in New Zealand. In Staufen im Breisgau, Germany, tectonic uplift occurred instead, due to a previously isolated anhydrite layer coming in contact with water and turning into gypsum, doubling its volume. Enhanced geothermal systems can trigger earthquakes as part of hydraulic fracturing. The project in Basel, Switzerland was suspended because more than 10,000 seismic events measuring up to 3.4 on the Richter Scale occurred over the first 6 days of water injection.

Geothermal has minimal land and freshwater requirements. Geothermal plants use 3.5 square kilometres (1.4 sq mi) per gigawatt of electrical production (not capacity) versus 32 square kilometres (12 sq mi) and 12 square kilometres (4.6 sq mi) for coal facilities and wind farms respectively. They use 20 litres (5.3 US gal) of freshwater per MW·h versus over 1,000 litres (260 US gal) per MW·h for nuclear, coal, or oil.

Legal Frameworks

Some of the legal issues raised by geothermal energy resources include questions of ownership and allocation of the resource, the grant of exploration permits, exploitation rights, royalties, and the extent to which geothermal energy issues have been recognized in existing planning and environmental laws. Other questions concern overlap between geothermal and mineral or petroleum tenements. Broader issues concern the extent to which the legal framework for encouragement of renewable energy assists in encouraging geothermal industry innovation and development.

Biomass

Sugarcane plantation in Brazil. Sugarcane bagasse is a type of biomass.

A cogeneration plant in Metz, France. The station uses waste wood biomass as an energy source, and provides electricity and heat for 30,000 dwellings.

Stump harvesting increases the recovery of biomass from a forest.

Biomass is organic matter derived from living, or recently living organisms. Biomass can be used as a source of energy and it most often refers to plants or plant-based materials which are not used for food or feed, and are specifically called lignocellulosic biomass. As an energy source, biomass can either be used directly via combustion to produce heat, or indirectly after converting it to various forms of biofuel. Conversion of biomass to biofuel can be achieved by different methods which are broadly classified into: *thermal*, *chemical*, and *biochemical* methods.

Biomass sources

Eucalyptus in Brazil. Remains of the tree are reused for power generation.

Historically, humans have harnessed biomass-derived energy since the time when people began burning wood to make fire. Even today, biomass is the only source of fuel for domestic use in many developing countries. Biomass is all biologically-produced matter based in carbon, hydrogen and oxygen. The estimated biomass production in the world is 104.9 petagrams (104.9 * 10^{15} g - about 105 billion metric tons) of carbon per year, about half in the ocean and half on land.

Wood remains the largest biomass energy source today; examples include forest residues (such as dead trees, branches and tree stumps), yard clippings, wood chips and even municipal solid waste. Wood energy is derived by using lignocellulosic biomass (second-generation biofuels) as fuel. Harvested wood may be used directly as a fuel or collected from wood waste streams. The largest source of energy from wood is pulping liquor or "black liquor," a waste product from processes of the pulp, paper and paperboard industry. In the second sense, biomass includes plant or animal matter that can be converted into fibers or other industrial chemicals, including biofuels. Industrial biomass can be grown from numerous types of plants, including miscanthus, switchgrass, hemp, corn, poplar, willow, sorghum, sugarcane, bamboo, and a variety of tree species, ranging from eucalyptus to oil palm (palm oil).

Based on the source of biomass, biofuels are classified broadly into two major categories. First-generation biofuels are derived from sources such as sugarcane and corn starch. Sugars present in this biomass are fermented to produce bioethanol, an alcohol fuel which can be used directly in a fuel cell to produce electricity or serve as an additive to gasoline. However, utilizing food-based resources for fuel production only aggravates the food shortage problem. Second-generation biofuels, on the other hand, utilize non-food-based biomass sources such as agriculture and municipal waste. These biofuels mostly consist of lignocellulosic biomass, which is not edible and is a low-value waste for many industries. Despite being the favored alternative, economical production of second-generation biofuel is not yet achieved due to technological issues. These issues arise mainly due to chemical inertness and structural rigidity of lignocellulosic biomass.

Plant energy is produced by crops specifically grown for use as fuel that offer high biomass output per hectare with low input energy. Some examples of these plants are wheat, which typically yields 7.5–8 tonnes of grain per hectare, and straw, which typically yields 3.5–5 tonnes per hectare in the UK. The grain can be used for liquid transportation fuels while the straw can be burned to produce heat or electricity. Plant biomass can also be degraded from cellulose to glucose through a series of chemical treatments, and the resulting sugar can then be used as a first-generation biofuel.

The main contributors of waste energy are municipal solid waste, manufacturing waste, and landfill gas. Energy derived from biomass is projected to be the largest non-hydroelectric renewable resource of electricity in the US between 2000 and 2020.

Biomass can be converted to other usable forms of energy like methane gas or transportation fuels like ethanol and biodiesel. Rotting garbage, and agricultural and human waste, all release methane gas, also called landfill gas or biogas. Crops such as corn and sugarcane can be fermented to produce the transportation fuel ethanol. Biodiesel, another transportation fuel, can be produced from leftover food products like vegetable oils and animal fats. Also, biomass-to-liquids (called "BTLs") and cellulosic ethanol are still under research.

There is research involving algae, or algae-derived, biomass, as this non-food resource can be produced at rates five to ten times those of other types of land-based agriculture, such as corn and soy. Once harvested, it can be fermented to produce biofuels such as ethanol, butanol, and methane, as well as biodiesel and hydrogen. Efforts are being made to identify which species of algae are most suitable for energy production. Genetic engineering approaches could also be utilized to improve microalgae as a source of biofuel.

The biomass used for electricity generation varies by region. Forest by-products, such as wood residues, are common in the US. Agricultural waste is common in Mauritius (sugar cane residue) and Southeast Asia (rice husks). Animal husbandry residues, such as poultry litter, are common in the UK.

As of 2015, a new bioenergy sewage treatment process aimed at developing countries is under trial; the Omni Processor is a self-sustaining process which uses sewerage solids as fuel in a process to convert waste water into drinking water, with surplus electrical energy being generated for export.

Comparison of Total Plant Biomass Yields (Dry Basis)

World Resources

If the total annual primary production of biomass is just over 100 billion (1.0E+11) tonnes of Carbon /yr, and the energy reserve per metric tonne of biomass is between about 1.5E3 – 3E3 Kilowatt hours (5E6 – 10E6 BTU), or 24.8 TW average, then biomass could in principle provide 1.4 times the approximate annual 150E3 Terrawatt.hours required for the current world energy consumption. For reference, the total solar power on Earth is 174 kTW. The biomass equivalent to solar energy ratio is 143 ppm (parts per million), given current living system coverage on Earth. Best in class solar cell efficiency is (20-40)%. Additionally, Earth's internal radioactive energy production, largely the driver for volcanic activity, continental drift, etc., is in the same range of power, 20 TW. At some 50% carbon mass content in biomass, annual production, this corresponds to about 6% of atmospheric carbon content in CO_2 (for the current 400 ppm).

(1.0E+11 tonnes biomass annually produced approximately 25 TW)

Annual world biomass energy equivalent =16.7 - 33.4 TW.

Annual world energy consumption =17.7. On average, biomass production is 1.4 times larger than world energy consumption.

Common Commodity Food Crops

- Agave: 1–21 tons/acre

- Alfalfa: 4–6 tons/acre

- Barley: grains – 1.6–2.8 tons/acre, straw – 0.9–2.5 tons/acre, total – 2.5–5.3 tons/acre

- Corn: grains – 3.2–4.9 tons/acre, stalks and stovers – 2.3–3.4 tons/acre, total – 5.5–8.3 tons/acre

- Jerusalem artichokes: tubers 1–8 tons/acre, tops 2–13 tons/acre, total 9–13 tons/acre

- Oats: grains – 1.4–5.4 tons/acre, straw – 1.9–3.2 tons/acre, total – 3.3–8.6 tons/acre

- Rye: grains – 2.1–2.4 tons/acre, straw – 2.4–3.4 tons/acre, total – 4.5–5.8 tons/acre

- Wheat: grains – 1.2–4.1 tons/acre, straw – 1.6–3.8 tons/acre, total – 2.8–7.9 tons/acre

Woody Crops

- Oil palm: fronds 11 ton/acre, whole fruit bunches 1 ton/acre, trunks 30 ton/acre

Not Yet in Commercial Planting

- Giant miscanthus: 5–15 tons/acre

- Sunn hemp: 4.5 tons/acre

- Switchgrass: 4–6 tons/acre

Genetically Modified Varieties

- Energy Sorghum

Biomass Conversion

Thermal Conversion

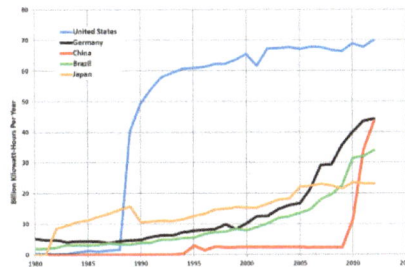

Trends in the top five countries generating electricity from biomass

Biomass briquettes are an example fuel for production of dendrothermal energy

Thermal conversion processes use heat as the dominant mechanism to convert biomass into another chemical form. The basic alternatives of combustion (torrefaction, pyrolysis, and gasification) are separated principally by the extent to which the chemical reactions involved are allowed to proceed (mainly controlled by the availability of oxygen and conversion temperature).

Energy created by burning biomass (fuel wood) is particularly suited for countries where the fuel wood grows more rapidly, e.g. tropical countries. There are a number of other less common, more experimental or proprietary thermal processes that may offer benefits such as hydrothermal upgrading (HTU) and hydroprocessing. Some have been developed for use on high moisture content biomass, including aqueous slurries, and allow them to be converted into more convenient forms. Some of the applications of thermal conversion are combined heat and power (CHP) and co-firing. In a typical dedicated biomass power plant, efficiencies range from 20–27% (higher heating value basis). Biomass cofiring with coal, by contrast, typically occurs at efficiencies near those of the coal combustor (30–40%, higher heating value basis).

Chemical Conversion

A range of chemical processes may be used to convert biomass into other forms, such as to produce a fuel that is more conveniently used, transported or stored, or to exploit some property of the process itself. Many of these processes are based in large part on similar coal-based processes, such as Fischer-Tropsch synthesis, methanol production, olefins (ethylene and propylene), and similar chemical or fuel feedstocks. In most cases, the first step involves gasification, which step generally is the most expensive and involves the greatest technical risk. Biomass is more difficult to feed into a pressure vessel than coal or any liquid. Therefore, biomass gasification is frequently done at atmospheric pressure and causes combustion of biomass to produce a combustible gas consisting of carbon monoxide, hydrogen, and traces of methane. This gas mixture, called a producer gas, can provide fuel for various vital processes, such as internal combustion engines, as well as substitute for furnace oil in direct heat applications. Because any biomass material can undergo gasification, this process is far more attractive than ethanol or biomass production, where only particular biomass materials can be used to produce a fuel. In addition, biomass gasification is a desirable process due to the ease at which it can convert solid waste (such as wastes available on a farm) into producer gas, which is a very usable fuel.

Conversion of biomass to biofuel can also be achieved via selective conversion of individual components of biomass. For example, cellulose can be converted to intermediate platform chemical such a sorbitol, glucose, hydroxymethylfurfural etc. These chemical are then further reacted to produce hydrogen or hydrocarbon fuels.

Biomass also has the potential to be converted to multiple commodity chemicals. Halomethanes have successfully been by produced using a combination of A. fermentans and engineered S. cerevisiae. This method converts NaX salts and unprocessed biomass such as switchgrass, sugarcane, corn stover, or poplar into halomethanes. S-adenosylmethionine which is naturally occurring in S. cerevisiae allows a methyl group to be transferred. Production levels of 150 mg $L-1H-1$ iodomethane were achieved. At these levels roughly 173000L of capacity would need to be operated just to replace the United States' need for iodomethane. However, an advantage of this method is that it uses NaI rather than I2; NaI is significantly less hazardous than I2. This method may be applied to produce ethylene in the future.

Other chemical processes such as converting straight and waste vegetable oils into biodiesel is transesterification.

Biochemical Conversion

As biomass is a natural material, many highly efficient biochemical processes have developed in nature to break down the molecules of which biomass is composed, and many of these biochemical conversion processes can be harnessed.

Biochemical conversion makes use of the enzymes of bacteria and other microorganisms to break down biomass. In most cases, microorganisms are used to perform the conversion process: anaerobic digestion, fermentation, and composting.

Electrochemical Conversion

In addition to combustion, bio-mass/bio-fuels can be directly converted to electrical energy via electrochemical oxidation of the material. This can be performed directly in a direct carbon fuel cell, direct ethanol fuel cell or a microbial fuel cell. The fuel can also be consumed indirectly via a fuel cell system containing a reformer which converts the bio-mass into a mixture of CO and H2 before it is consumed in the fuel cell.

In the United States

The biomass power generating industry in the United States, which consists of approximately 11,000 MW of summer operating capacity actively supplying power to the grid, produces about 1.4 percent of the U.S. electricity supply.

Currently, the New Hope Power Partnership is the largest biomass power plant in the U.S. The 140 MW facility uses sugarcane fiber (bagasse) and recycled urban wood as fuel to generate enough power for its large milling and refining operations as well as to supply electricity for nearly 60,000 homes.

Second-generation Biofuels

Second-generation biofuels were not (in 2010) produced commercially, but a considerable number of research activities were taking place mainly in North America, Europe and also in some emerging countries. These tend to use feedstock produced by rapidly reproducing enzymes or bacteria from various sources including excrement grown in Cell cultures or hydroponics There is huge potential for second generation biofuels but non-edible feedstock resources are highly underutilized.

Environmental Impact

Using biomass as a fuel produces air pollution in the form of carbon monoxide, carbon dioxide, NOx (nitrogen oxides), VOCs (volatile organic compounds), particulates and other pollutants at levels above those from traditional fuel sources such as coal or natural gas in some cases (such as with indoor heating and cooking). Utilization of wood biomass as a fuel can also produce fewer particulate and other pollutants than open burning as seen in wildfires or direct heat applications.

Black carbon – a pollutant created by combustion of fossil fuels, biofuels, and biomass – is possibly the second largest contributor to global warming. In 2009 a Swedish study of the giant brown haze that periodically covers large areas in South Asia determined that it had been principally produced by biomass burning, and to a lesser extent by fossil-fuel burning. Researchers measured a significant concentration of ^{14}C, which is associated with recent plant life rather than with fossil fuels.

Biomass power plant size is often driven by biomass availability in close proximity as transport costs of the (bulky) fuel play a key factor in the plant's economics. It has to be noted, however, that rail and especially shipping on waterways can reduce transport costs significantly, which has led to a global biomass market. To make small plants of 1 MW_{el} economically profitable those power plants need to be equipped with technology that is able to convert biomass to useful electricity with high efficiency such as ORC technology, a cycle similar to the water steam power process just with an organic working medium. Such small power plants can be found in Europe.

On combustion, the carbon from biomass is released into the atmosphere as carbon dioxide (CO_2). The amount of carbon stored in dry wood is approximately 50% by weight. However, according to the Food and Agriculture Organization of the United Nations, plant matter used as a fuel can be replaced by planting for new growth. When the biomass is from forests, the time to recapture the carbon stored is generally longer, and the carbon storage capacity of the forest may be reduced overall if destructive forestry techniques are employed.

Industry professionals claim that a range of issues can affect a plant's ability to comply with emissions standards. Some of these challenges, unique to biomass plants, include inconsistent fuel supplies and age. The type and amount of the fuel supply are completely reliant factors; the fuel can be in the form of building debris or agricultural waste (such as deforestation of invasive species or orchard trimmings). Furthermore, many of the biomass plants are old, use outdated technology and were not built to comply with today's stringent standards. In fact, many are based on technologies developed during the term of U.S. President Jimmy Carter, who created the United States Department of Energy in 1977.

The U.S. Energy Information Administration projected that by 2017, biomass is expected to be about twice as expensive as natural gas, slightly more expensive than nuclear power, and much less expensive than solar panels. In another EIA study released, concerning the government's plan to implement a 25% renewable energy standard by 2025, the agency assumed that 598 million tons of biomass would be available, accounting for 12% of the renewable energy in the plan.

The adoption of biomass-based energy plants has been a slow but steady process. Between the years of 2002 and 2012 the production of these plants has increased 14%. In the United States, alternative electricity-production sources on the whole generate about 13% of power; of this fraction, biomass contributes approximately 11% of the alternative production. According to a study conducted in early 2012, of the 107 operating biomass plants in the United States, 85 have been cited by federal or state regulators for the violation of clean air or water standards laws over the past 5 years. This data also includes minor infractions.

Despite harvesting, biomass crops may sequester carbon. For example, soil organic carbon has been observed to be greater in switchgrass stands than in cultivated cropland soil, especially at depths below 12 inches. The grass sequesters the carbon in its increased root biomass. Typically, perennial crops sequester much more carbon than annual crops due to much greater non-harvested living biomass, both living and dead, built up over years, and much less soil disruption in cultivation.

The proposal that biomass is carbon-neutral put forward in the early 1990s has been superseded by more recent science that recognizes that mature, intact forests sequester carbon more effectively than cut-over areas. When a tree's carbon is released into the atmosphere in a single pulse, it contributes to climate change much more than woodland timber rotting slowly over decades. Current studies indicate that "even after 50 years the forest has not recovered to its initial carbon storage" and "the optimal strategy is likely to be protection of the standing forest".

The pros and cons of biomass usage regarding carbon emissions may be quantified with the ILUC factor. There is controversy surrounding the usage of the ILUC factor.

Forest-based biomass has recently come under fire from a number of environmental organizations, including Greenpeace and the Natural Resources Defense Council, for the harmful impacts it can have on forests and the climate. Greenpeace recently released a report entitled "Fuelling a BioMess" which outlines their concerns around forest-based biomass. Because any part of the tree can be burned, the harvesting of trees for energy production encourages Whole-Tree Harvesting, which removes more nutrients and soil cover than regular harvesting, and can be harmful to the long-term health of the forest. In some jurisdictions, forest biomass removal is increasingly involving elements essential to functioning forest ecosystems, including standing trees, naturally disturbed forests and remains of traditional logging operations that were previously left in the forest. Environmental groups also cite recent scientific research which has found that it can take many decades for the carbon released by burning biomass to be recaptured by regrowing trees, and even longer in low productivity areas; furthermore, logging operations may disturb forest soils and cause them to release stored carbon. In light of the pressing need to reduce greenhouse gas emissions in the short term in order to mitigate the effects of climate change, a number of environmental groups are opposing the large-scale use of forest biomass in energy production.

Supply Chain Issues

With the seasonality of biomass supply and a great variability in sources, supply chains play a key role in cost-effective delivery of bioenergy. There are several potential challenges peculiar to bioenergy supply chains:

Technical issues

- Inefficiencies of conversion

- Storage methods for seasonal availability

- Complex multi-component constituents incompatible with maximizing efficiency of single purpose use

- High water content

- Conflicting decisions (technologies, locations, and routes)

- Complex location analysis (source points, inventory facilities, and production plants)

Logistic issues

- Seasonal availability leading to storage challenges and/or seasonally idle facilities
- Low bulk-density and/or high water content
- Finite productivity per area and/or time incompatible with conventional approach to economy of scale focusing on maximizing facility size

Financial issues

- The limits for the traditional approach to economy of scale which focuses on maximizing single facility size
- Unavailability and complexity of life cycle costing data
- Lack of required transport infrastructure
- Limited flexibility or inflexibility to energy demand
- Risks associated with new technologies (insurability, performance, rate of return)
- Extended market volatilities (conflicts with alternative markets for biomass)
- Difficult or impossible to use financial hedging methods to control cost

Social issues

- Lack of participatory decision making
- Lack of public/community awareness
- Local supply chain impacts vs. global benefits
- Health and safety risks
- Extra pressure on transport sector
- Decreasing the esthetics of rural areas

Policy and regulatory issues

- Impact of fossil fuel tax on biomass transport
- Lack of incentives to create competition among bioenergy producers
- Focus on technology options and less attention to selection of biomass materials
- Lack of support for sustainable supply chain solutions

Institutional and organizational issues

- Varied ownership arrangements and priorities among supply chain parties
- Lack of supply chain standards
- Impact of organizational norms and rules on decision making and supply chain coordination
- Immaturity of change management practices in biomass supply chains

Biofuel

A bus fueled by biodiesel

Information on pump regarding ethanol fuel blend up to 10%, California

A biofuel is a fuel that is produced through contemporary biological processes, such as agriculture and anaerobic digestion, rather than a fuel produced by geological processes such as those involved in the formation of fossil fuels, such as coal and petroleum, from prehistoric biological matter. Biofuels can be derived directly from plants, or indirectly from agricultural, commercial, domestic, and/or industrial wastes. Renewable biofuels generally involve contemporary carbon fixation, such as those that occur in plants or microalgae through the process of photosynthesis. Other renewable biofuels are made through the use or conversion of biomass (referring to recently living organisms, most often referring to plants or plant-derived materials). This biomass can be converted to convenient energy-containing substances in three different ways: thermal conversion, chemical conversion, and biochemical conversion. This biomass conversion can result in fuel in solid, liquid, or gas form. This new biomass can also be used directly for biofuels.

Bioethanol is an alcohol made by fermentation, mostly from carbohydrates produced in sugar or starch crops such as corn, sugarcane, or sweet sorghum. Cellulosic biomass, derived from non-food sources, such as trees and grasses, is also being developed as a feedstock for ethanol production. Ethanol can be used as a fuel for vehicles in its pure form, but it is usually used as a gasoline additive to increase octane and improve vehicle emissions. Bioethanol is widely used in the USA and in Brazil. Current plant design does not provide for converting the lignin portion of plant raw materials to fuel components by fermentation.

Biodiesel can be used as a fuel for vehicles in its pure form, but it is usually used as a diesel additive to reduce levels of particulates, carbon monoxide, and hydrocarbons from diesel-powered vehicles. Biodiesel is produced from oils or fats using transesterification and is the most common biofuel in Europe.

In 2010, worldwide biofuel production reached 105 billion liters (28 billion gallons US), up 17% from 2009, and biofuels provided 2.7% of the world's fuels for road transport. Global ethanol fuel production reached 86 billion liters (23 billion gallons US) in 2010, with the United States and Brazil as the world's top producers, accounting together for 90% of global production. The world's largest biodiesel producer is the European Union, accounting for 53% of all biodiesel production in 2010. As of 2011, mandates for blending biofuels exist in 31 countries at the national level and in 29 states or provinces. The International Energy Agency has a goal for biofuels to meet more than a quarter of world demand for transportation fuels by 2050 to reduce dependence on petroleum and coal. The production of biofuels also led into a flourishing automotive industry, where by 2010, 79% of all cars produced in Brazil were made with a hybrid fuel system of bioethanol and gasoline.

There are various social, economic, environmental and technical issues relating to biofuels production and use, which have been debated in the popular media and scientific journals. These include: the effect of moderating oil prices, the "food vs fuel" debate, poverty reduction potential, carbon emissions levels, sustainable biofuel production, deforestation and soil erosion, loss of biodiversity, impact on water resources, rural social exclusion and injustice, shantytown migration, rural unskilled unemployment, and nitrous oxide (NO_2) emissions.

Liquid Fuels for Transportation

Most transportation fuels are liquids, because vehicles usually require high energy density. This occurs naturally in liquids and solids. High energy density can also be provided by an internal combustion engine. These engines require clean-burning fuels. The fuels that are easiest to burn cleanly are typically liquids and gases. Thus, liquids meet the requirements of being both energy-dense and clean-burning. In addition, liquids (and gases) can be pumped, which means handling is easily mechanized, and thus less laborious.

First-generation Biofuels

"First-generation" or conventional biofuels are made from sugar, starch, or vegetable oil.

Ethanol

Biologically produced alcohols, most commonly ethanol, and less commonly propanol and butanol, are produced by the action of microorganisms and enzymes through the fermentation of sugars or starches (easiest), or cellulose (which is more difficult). Biobutanol (also called biogasoline) is often claimed to provide a direct replacement for gasoline, because it can be used directly in a gasoline engine.

Neat ethanol on the left (A), gasoline on the right (G) at a filling station in Brazil

Ethanol fuel is the most common biofuel worldwide, particularly in Brazil. Alcohol fuels are produced by fermentation of sugars derived from wheat, corn, sugar beets, sugar cane, molasses and any sugar or starch from which alcoholic beverages such as whiskey, can be made (such as potato and fruit waste, etc.). The ethanol production methods used are enzyme digestion (to release sugars from stored starches), fermentation of the sugars, distillation and drying. The distillation process requires significant energy input for heat (sometimes unsustainable natural gas fossil fuel, but cellulosic biomass such as bagasse, the waste left after sugar cane is pressed to extract its juice, is the most common fuel in Brazil, while pellets, wood chips and also waste heat are more common in Europe) Waste steam fuels ethanol factory- where waste heat from the factories also is used in the district heating grid.

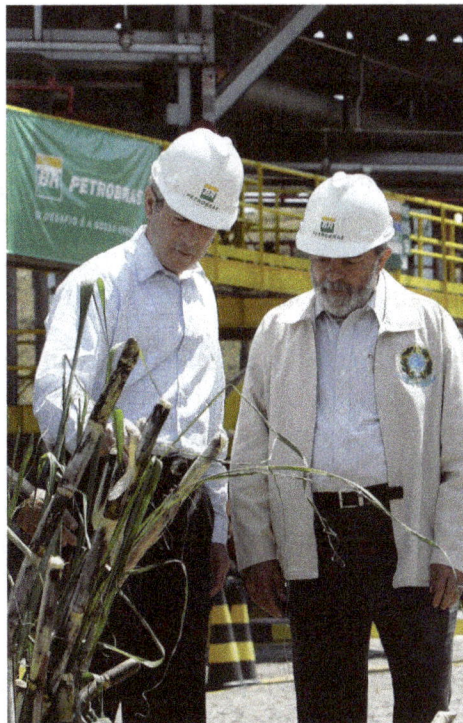

U.S. President George W. Bush looks at sugar cane, a source of biofuel, with Brazilian President Luiz Inácio Lula da Silva during a tour on biofuel technology at Petrobras in São Paulo, Brazil, 9 March 2007.

Ethanol can be used in petrol engines as a replacement for gasoline; it can be mixed with gasoline to any percentage. Most existing car petrol engines can run on blends of up to 15% bioethanol with petroleum/gasoline. Ethanol has a smaller energy density than that of gasoline; this means it takes more fuel (volume and mass) to produce the same amount of work. An advantage of ethanol (CH3CH2OH) is that it has a higher octane rating than ethanol-free gasoline available at roadside gas stations, which allows an increase of an engine's compression ratio for increased thermal efficiency. In high-altitude (thin air) locations, some states mandate a mix of gasoline and ethanol as a winter oxidizer to reduce atmospheric pollution emissions.

Ethanol is also used to fuel bioethanol fireplaces. As they do not require a chimney and are "flueless", bioethanol fires are extremely useful for newly built homes and apartments without a flue. The downsides to these fireplaces is that their heat output is slightly less than electric heat or gas fires, and precautions must be taken to avoid carbon monoxide poisoning.

Corn-to-ethanol and other food stocks has led to the development of cellulosic ethanol. According to a joint research agenda conducted through the US Department of Energy, the fossil energy ratios (FER) for cellulosic ethanol, corn ethanol, and gasoline are 10.3, 1.36, and 0.81, respectively.

Ethanol has roughly one-third lower energy content per unit of volume compared to gasoline. This is partly counteracted by the better efficiency when using ethanol (in a long-term test of more than 2.1 million km, the BEST project found FFV vehicles to be 1-26 % more energy efficient than petrol cars The BEST project), but the volumetric consumption increases by approximately 30%, so more fuel stops are required.

With current subsidies, ethanol fuel is slightly cheaper per distance traveled in the United States.

Biodiesel

Biodiesel is the most common biofuel in Europe. It is produced from oils or fats using transesterification and is a liquid similar in composition to fossil/mineral diesel. Chemically, it consists mostly of fatty acid methyl (or ethyl) esters (FAMEs). Feedstocks for biodiesel include animal fats, vegetable oils, soy, rapeseed, jatropha, mahua, mustard, flax, sunflower, palm oil, hemp, field pennycress, *Pongamia pinnata* and algae. Pure biodiesel (B100) currently reduces emissions with up to 60% compared to diesel Second generation B100.

Biodiesel can be used in any diesel engine when mixed with mineral diesel. In some countries, manufacturers cover their diesel engines under warranty for B100 use, although Volkswagen of Germany, for example, asks drivers to check by telephone with the VW environmental services department before switching to B100. B100 may become more viscous at lower temperatures, depending on the feedstock used. In most cases, biodiesel is compatible with diesel engines from 1994 onwards, which use 'Viton' (by DuPont) synthetic rubber in their mechanical fuel injection systems. Note however, that no vehicles are certified for using neat biodiesel before 2014, as there was no emission control protocol available for biodiesel before this date.

Electronically controlled 'common rail' and 'unit injector' type systems from the late 1990s onwards may only use biodiesel blended with conventional diesel fuel. These engines have finely metered and atomized multiple-stage injection systems that are very sensitive to the viscosity of the fuel. Many current-generation diesel engines are made so that they can run on B100 without altering the engine itself, although this depends on the fuel rail design. Since biodiesel is an effective solvent and cleans residues deposited by mineral diesel, engine filters may need to be replaced more often, as the biofuel dissolves old deposits in the fuel tank and pipes. It also effectively cleans the engine combustion chamber of carbon deposits, helping to maintain efficiency. In many European countries, a 5% biodiesel blend is widely used and is available at thousands of gas stations. Biodiesel is also an oxygenated fuel, meaning it contains a reduced amount of carbon and higher hydrogen and oxygen content than fossil diesel. This improves the combustion of biodiesel and reduces the particulate emissions from unburnt carbon. However, using neat biodiesel may increase NOx-emissions Nylund.N-O & Koponen.K. 2013. Fuel and Technology Alternatives for Buses. Overall Energy Efficiency and Emission Performance. IEA Bioenergy Task 46. Possibly the new emission standards Euro VI/EPA 10 will lead to reduced NOx-levels also when using B100.

Biodiesel is also safe to handle and transport because it is non-toxic and biodegradable, and has a high flash point of about 300 °F (148 °C) compared to petroleum diesel fuel, which has a flash point of 125 °F (52 °C).

In the USA, more than 80% of commercial trucks and city buses run on diesel. The emerging US biodiesel market is estimated to have grown 200% from 2004 to 2005. "By the end of 2006 biodiesel production was estimated to increase fourfold [from 2004] to more than" 1 billion US gallons (3,800,000 m³).

In France, biodiesel is incorporated at a rate of 8% in the fuel used by all French diesel vehicles. Avril Group produces under the brand Diester, a fifth of 11 million tons of biodiesel consumed annually by the European Union. It is the leading European producer of biodiesel.

Other Bioalcohols

Methanol is currently produced from natural gas, a non-renewable fossil fuel. In the future it is hoped to be produced from biomass as biomethanol. This is technically feasible, but the production is currently being postponed for concerns of Jacob S. Gibbs and Brinsley Coleberd that the economic viability is still pending. The methanol economy is an alternative to the hydrogen economy, compared to today's hydrogen production from natural gas.

Butanol (C 4H 9OH) is formed by ABE fermentation (acetone, butanol, ethanol) and experimental modifications of the process show potentially high net energy gains with butanol as the only liquid product. Butanol will produce more energy and allegedly can be burned "straight" in existing gasoline engines (without modification to the engine or car), and is less corrosive and less water-soluble than ethanol, and could be distributed via existing infrastructures. DuPont and BP are working together to help develop butanol. *E. coli* strains have also been successfully engineered to produce butanol by modifying their amino acid metabolism.

Green Diesel

Green diesel is produced through hydrocracking biological oil feedstocks, such as vegetable oils and animal fats. Hydrocracking is a refinery method that uses elevated temperatures and pressure in the presence of a catalyst to break down larger molecules, such as those found in vegetable oils, into shorter hydrocarbon chains used in diesel engines. It may also be called renewable diesel, hydrotreated vegetable oil or hydrogen-derived renewable diesel. Green diesel has the same chemical properties as petroleum-based diesel. It does not require new engines, pipelines or infrastructure to distribute and use, but has not been produced at a cost that is competitive with petroleum. Gasoline versions are also being developed. Green diesel is being developed in Louisiana and Singapore by ConocoPhillips, Neste Oil, Valero, Dynamic Fuels, and Honeywell UOP as well as Preem in Gothenburg, Sweden, creating what is known as Evolution Diesel.

Biofuel Gasoline

In 2013 UK researchers developed a genetically modified strain of Escherichia coli (E.Coli), which could transform glucose into biofuel gasoline that does not need to be blended. Later in 2013 UCLA researchers engineered a new metabolic pathway to bypass glycolysis and increase the rate of conversion of sugars into biofuel, while KAIST researchers developed a strain capable of producing short-chain alkanes, free fatty acids, fatty esters and fatty alcohols through the fatty acyl (acyl carrier protein (ACP)) to fatty acid to fatty acyl-CoA pathway *in vivo*. It is believed that in the future it will be possible to "tweak" the genes to make gasoline from straw or animal manure.

Vegetable Oil

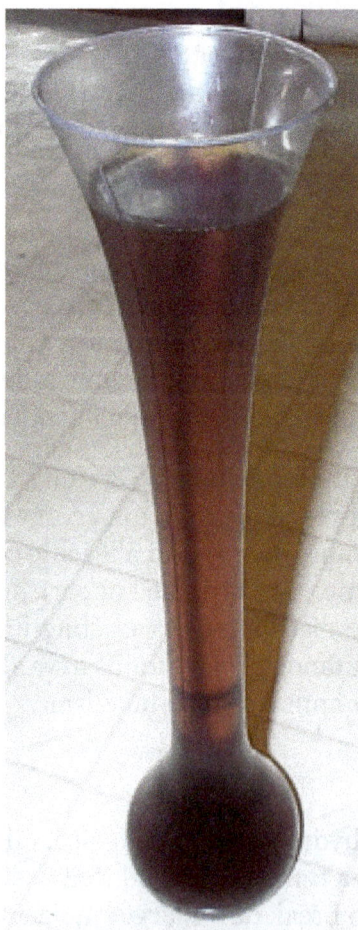

Filtered waste vegetable oil

Walmart's truck fleet logs millions of miles each year, and the company planned to double the fleet's efficiency between 2005 and 2015. This truck is one of 15 based at Walmart's Buckeye, Arizona distribution center that was converted to run on a biofuel made from reclaimed cooking grease produced during food preparation at Walmart stores.

Straight unmodified edible vegetable oil is generally not used as fuel, but lower-quality oil can and has been used for this purpose. Used vegetable oil is increasingly being processed into biodiesel, or (more rarely) cleaned of water and particulates and used as a fuel.

As with 100% biodiesel (B100), to ensure the fuel injectors atomize the vegetable oil in the correct pattern for efficient combustion, vegetable oil fuel must be heated to reduce its viscosity to that of diesel, either by electric coils or heat exchangers. This is easier in warm or temperate climates. MAN B&W Diesel, Wärtsilä, and Deutz AG, as well as a number of smaller companies, such as Elsbett, offer engines that are compatible with straight vegetable oil, without the need for after-market modifications.

Vegetable oil can also be used in many older diesel engines that do not use common rail or unit injection electronic diesel injection systems. Due to the design of the combustion chambers in indirect injection engines, these are the best engines for use with vegetable oil. This system allows the relatively larger oil molecules more time to burn. Some older engines, especially Mercedes, are driven experimentally by enthusiasts without any conversion, a handful of drivers have experienced limited success with earlier pre-"Pumpe Duse" VW TDI engines and other similar engines with direct injection. Several companies, such as Elsbett or Wolf, have developed professional conversion kits and successfully installed hundreds of them over the last decades.

Oils and fats can be hydrogenated to give a diesel substitute. The resulting product is a straight-chain hydrocarbon with a high cetane number, low in aromatics and sulfur and does not contain oxygen. Hydrogenated oils can be blended with diesel in all proportions. They have several advantages over biodiesel, including good performance at low temperatures, no storage stability problems and no susceptibility to microbial attack.

Bioethers

Bioethers (also referred to as fuel ethers or oxygenated fuels) are cost-effective compounds that act as octane rating enhancers."Bioethers are produced by the reaction of reactive iso-olefins, such as iso-butylene, with bioethanol." Bioethers are created by wheat or sugar beet. They also enhance engine performance, whilst significantly reducing engine wear and toxic exhaust emissions. Though bioethers are likely to replace petroethers in the UK, it is highly unlikely they will become a fuel in and of itself due to the low energy density. Greatly reducing the amount of ground-level ozone emissions, they contribute to air quality.

When it comes to transportation fuel there are six ether additives- 1. Dimethyl Ether (DME) 2. Diethyl Ether (DEE) 3. Methyl Teritiary-Butyl Ether (MTBE) 4. Ethyl *ter*-butyl ether (ETBE) 5. *Ter*-amyl methyl ether (TAME) 6. *Ter*-amyl ethyl Ether (TAEE)

The European Fuel Oxygenates Association (aka EFOA) credits Methyl Tertiary-Butyl Ether (MTBE) and Ethyl ter-butyl ether (ETBE) as the most commonly used ethers in fuel to replace lead. Ethers were brought into fuels in Europe in the 1970s to replace the highly toxic compound. Although Europeans still use Bio-ether additives, the US no longer has an oxygenate requirement therefore bio-ethers are no longer used as the main fuel additive.

Biogas

Pipes carrying biogas

Biogas is methane produced by the process of anaerobic digestion of organic material by anaerobes. It can be produced either from biodegradable waste materials or by the use of energy crops fed into anaerobic digesters to supplement gas yields. The solid byproduct, digestate, can be used as a biofuel or a fertilizer.

• Biogas can be recovered from mechanical biological treatment waste processing systems.

Note: Landfill gas, a less clean form of biogas, is produced in landfills through naturally occurring anaerobic digestion. If it escapes into the atmosphere, it is a potential greenhouse gas.

• Farmers can produce biogas from manure from their cattle by using anaerobic digesters.

Syngas

Syngas, a mixture of carbon monoxide, hydrogen and other hydrocarbons, is produced by partial combustion of biomass, that is, combustion with an amount of oxygen that is not sufficient to convert the biomass completely to carbon dioxide and water. Before partial combustion, the biomass is dried, and sometimes pyrolysed. The resulting gas mixture, syngas, is more efficient than direct combustion of the original biofuel; more of the energy contained in the fuel is extracted.

- Syngas may be burned directly in internal combustion engines, turbines or high-temperature fuel cells. The wood gas generator, a wood-fueled gasification reactor, can be connected to an internal combustion engine.

- Syngas can be used to produce methanol, DME and hydrogen, or converted via the Fischer-Tropsch process to produce a diesel substitute, or a mixture of alcohols that can be blended into gasoline. Gasification normally relies on temperatures greater than 700 °C.

- Lower-temperature gasification is desirable when co-producing biochar, but results in syngas polluted with tar.

Solid Biofuels

Examples include wood, sawdust, grass trimmings, domestic refuse, charcoal, agricultural waste, nonfood energy crops, and dried manure.

When raw biomass is already in a suitable form (such as firewood), it can burn directly in a stove or furnace to provide heat or raise steam. When raw biomass is in an inconvenient form (such as sawdust, wood chips, grass, urban waste wood, agricultural residues), the typical process is to densify the biomass. This process includes grinding the raw biomass to an appropriate particulate size (known as hogfuel), which, depending on the densification type, can be from 1 to 3 cm (0.4 to 1.2 in), which is then concentrated into a fuel product. The current processes produce wood pellets, cubes, or pucks. The pellet process is most common in Europe, and is typically a pure wood product. The other types of densification are larger in size compared to a pellet, and are compatible with a broad range of input feedstocks. The resulting densified fuel is easier to transport and feed into thermal generation systems, such as boilers.

Industry has used sawdust, bark and chips for fuel for decades, primary in the pulp and paper industry, and also bagasse (spent sugar cane) fueled boilers in the sugar cane industry. Boilers in the range of 500,000 lb/hr of steam, and larger, are in routine operation, using grate, spreader stoker, suspension burning and fluid bed combustion. Utilities generate power, typically in the range of 5 to 50 MW, using locally available fuel. Other industries have also installed wood waste fueled boilers and dryers in areas with low cost fuel.

One of the advantages of biomass fuel is that it is often a byproduct, residue or waste-product of other processes, such as farming, animal husbandry and forestry. In theory, this means fuel and food production do not compete for resources, although this is not always the case.

A problem with the combustion of raw biomass is that it emits considerable amounts of pollutants, such as particulates and polycyclic aromatic hydrocarbons. Even modern pellet boilers generate much more pollutants than oil or natural gas boilers. Pellets made from agricultural residues are usually worse than wood pellets, producing much larger emissions of dioxins and chlorophenols.

In spite of the above noted study, numerous studies have shown biomass fuels have significantly less impact on the environment than fossil based fuels. Of note is the US Department of Energy Laboratory, operated by Midwest Research Institute Biomass Power and Conventional Fossil Systems with and without CO2 Sequestration – Comparing the Energy Balance, Greenhouse Gas Emissions and Economics Study. Power generation emits significant amounts of greenhouse gases (GHGs), mainly carbon dioxide (CO_2). Sequestering CO_2 from the power plant flue gas can significantly reduce the GHGs from the power plant itself, but this is not the total picture. CO_2 capture and sequestration consumes additional energy, thus lowering the plant's fuel-to-electricity efficiency. To compensate for this, more fossil fuel must be procured and consumed to make up for lost capacity.

Taking this into consideration, the global warming potential (GWP), which is a combination of CO_2, methane (CH_4), and nitrous oxide (N_2O) emissions, and energy balance of the system need to be examined using a life cycle assessment. This takes into account the upstream processes which remain constant after CO_2 sequestration, as well as the steps required for additional power generation. Firing biomass instead of coal led to a 148% reduction in GWP.

A derivative of solid biofuel is biochar, which is produced by biomass pyrolysis. Biochar made from agricultural waste can substitute for wood charcoal. As wood stock becomes scarce, this alternative is gaining ground. In eastern Democratic Republic of Congo, for example, biomass briquettes are being marketed as an alternative to charcoal to protect Virunga National Park from deforestation associated with charcoal production.

Second-generation (Advanced) Biofuels

Second generation biofuels, also known as advanced biofuels, are fuels that can be manufactured from various types of biomass. Biomass is a wide-ranging term meaning any source of organic carbon that is renewed rapidly as part of the carbon cycle. Biomass is derived from plant materials but can also include animal materials.

First generation biofuels are made from the sugars and vegetable oils found in arable crops, which can be easily extracted using conventional technology. In comparison, second generation biofuels are made from lignocellulosic biomass or woody crops, agricultural residues or waste, which makes it harder to extract the required fuel. A series of physical and chemical treatments might be required to convert lignocellulosic biomass to liquid fuels suitable for transportation.

Sustainable Biofuels

Biofuels in the form of liquid fuels derived from plant materials, are entering the market, driven mainly by the perception that they reduce climate gas emissions, and also by factors such as oil price spikes and the need for increased energy security. However, many of the biofuels that are currently being supplied have been criticised for their adverse impacts on the natural environment, food security, and land use. In 2008, the Nobel-prize winning chemist Paul J. Crutzen published findings that the release of nitrous oxide (N_2O) emissions in the production of biofuels means that overall they contribute more to global warming than the fossil fuels they replace.

The challenge is to support biofuel development, including the development of new cellulosic technologies, with responsible policies and economic instruments to help ensure that biofuel commercialization is sustainable. Responsible commercialization of biofuels represents an opportunity to enhance sustainable economic prospects in Africa, Latin America and Asia.

According to the Rocky Mountain Institute, sound biofuel production practices would not hamper food and fibre production, nor cause water or environmental problems, and would enhance soil fertility. The selection of land on which to grow the feedstocks is a critical component of the ability of biofuels to deliver sustainable solutions. A key consideration is the minimisation of biofuel competition for prime cropland.

Biofuels by Region

Bio Diesel Powered Fast Attack Craft Of Indian Navy patrolling during IFR 2016.
The green bands on the vessels are indicative of the fact that the vessels are powered by bio-diesel

There are international organizations such as IEA Bioenergy, established in 1978 by the OECD International Energy Agency (IEA), with the aim of improving cooperation and information exchange between countries that have national programs in bioenergy research, development and deployment. The UN International Biofuels Forum is formed by Brazil, China, India, Pakistan, South Africa, the United States and the European Commission. The world leaders in biofuel development and use are Brazil, the United States, France, Sweden and Germany. Russia also has 22% of world's forest, and is a big biomass (solid biofuels) supplier. In 2010, Russian pulp and paper maker, Vyborgskaya Cellulose, said they would be producing pellets that can be used in heat and electricity generation from its plant in Vyborg by the end of the year. The plant will eventually produce about 900,000 tons of pellets per year, making it the largest in the world once operational.

Biofuels currently make up 3.1% of the total road transport fuel in the UK or 1,440 million litres. By 2020, 10% of the energy used in UK road and rail transport must come from renewable sources – this is the equivalent of replacing 4.3 million tonnes of fossil oil each year. Conventional biofuels are likely to produce between 3.7 and 6.6% of the energy needed in road and rail transport, while advanced biofuels could meet up to 4.3% of the UK's renewable transport fuel target by 2020.

Air Pollution

Biofuels are different from fossil fuels in regard to greenhouse gases but are similar to fossil fuels in that biofuels contribute to air pollution. Burning produces airborne carbon particulates, carbon monoxide and nitrous oxides. The WHO estimates 3.7 million premature deaths worldwide in 2012 due to air pollution. Brazil burns significant amounts of ethanol biofuel. Gas chromatograph studies were performed of ambient air in São Paulo, Brazil, and compared to Osaka, Japan, which does not burn ethanol fuel. Atmospheric Formaldehyde was 160% higher in Brazil, and Acetaldehyde was 260% higher.

Debates Regarding the Production and use of Biofuel

There are various social, economic, environmental and technical issues with biofuel production and use, which have been discussed in the popular media and scientific journals. These include: the effect of moderating oil prices, the "food vs fuel" debate, food prices, poverty reduction potential, energy ratio, energy requirements, carbon emissions levels, sustainable biofuel production, deforestation and soil erosion, loss of biodiversity, impact on water resources, the possible modifications necessary to run the engine on biofuel, as well as energy balance and efficiency. The International Resource Panel, which provides independent scientific assessments and expert advice on a variety of resource-related themes, assessed the issues relating to biofuel use in its first report *Towards sustainable production and use of resources: Assessing Biofuels*. "Assessing Biofuels" outlined the wider and interrelated factors that need to be considered when deciding on the relative merits of pursuing one biofuel over another. It concluded that not all biofuels perform equally in terms of their impact on climate, energy security and ecosystems, and suggested that environmental and social impacts need to be assessed throughout the entire life-cycle.

Another issue with biofuel use and production is the US has changed mandates many times because the production has been taking longer than expected. The Renewable Fuel Standard (RFS) set by congress for 2010 was pushed back to at best 2012 to produce 100 million gallons of pure ethanol (not blended with a fossil fuel).

Current Research

Research is ongoing into finding more suitable biofuel crops and improving the oil yields of these crops. Using the current yields, vast amounts of land and fresh water would be needed to produce enough oil to completely replace fossil fuel usage. It would require twice the land area of the US to be devoted to soybean production, or two-thirds to be devoted to rapeseed production, to meet current US heating and transportation needs.

Specially bred mustard varieties can produce reasonably high oil yields and are very useful in crop rotation with cereals, and have the added benefit that the meal left over after the oil has been pressed out can act as an effective and biodegradable pesticide.

The NFESC, with Santa Barbara-based Biodiesel Industries, is working to develop biofuels technologies for the US navy and military, one of the largest diesel fuel users in the world. A group of Spanish developers working for a company called Ecofasa announced a new biofuel made from trash. The fuel is created from general urban waste which is treated by bacteria to produce fatty acids, which can be used to make biofuels.

Ethanol Biofuels

As the primary source of biofuels in North America, many organizations are conducting research in the area of ethanol production. The National Corn-to-Ethanol Research Center (NCERC) is a research division of Southern Illinois University Edwardsville dedicated solely to ethanol-based biofuel research projects. On the federal level, the USDA conducts a large amount of research regarding ethanol production in the United States. Much of this research is targeted toward the effect of ethanol production on domestic food markets. A division of the U.S. Department of Energy, the National Renewable Energy Laboratory (NREL), has also conducted various ethanol research projects, mainly in the area of cellulosic ethanol.

Cellulosic ethanol commercialization is the process of building an industry out of methods of turning cellulose-containing organic matter into fuel. Companies, such as Iogen, POET, and Abengoa, are building refineries that can process biomass and turn it into bioethanol. Companies, such as Diversa, Novozymes, and Dyadic, are producing enzymes that could enable a cellulosic ethanol future. The shift from food crop feedstocks to waste residues and native grasses offers significant opportunities for a range of players, from farmers to biotechnology firms, and from project developers to investors.

As of 2013, the first commercial-scale plants to produce cellulosic biofuels have begun operating. Multiple pathways for the conversion of different biofuel feedstocks are being used. In the next few years, the cost data of these technologies operating at commercial scale, and their relative performance, will become available. Lessons learnt will lower the costs of the industrial processes involved.

In parts of Asia and Africa where drylands prevail, sweet sorghum is being investigated as a potential source of food, feed and fuel combined. The crop is particularly suitable for growing in arid conditions, as it only extracts one seventh of the water used by sugarcane. In India, and other places, sweet sorghum stalks are used to produce biofuel by squeezing the juice and then fermenting into ethanol.

A study by researchers at the International Crops Research Institute for the Semi-Arid Tropics (ICRISAT) found that growing sweet sorghum instead of grain sorghum could increase farmers incomes by US$40 per hectare per crop because it can provide fuel in addition to food and animal feed. With grain sorghum currently grown on over 11 million hectares (ha) in Asia and on 23.4 million ha in Africa, a switch to sweet sorghum could have a considerable economic impact.

Algae Biofuels

From 1978 to 1996, the US NREL experimented with using algae as a biofuels source in the "Aquatic Species Program". A self-published article by Michael Briggs, at the UNH Biofuels Group, offers estimates for the realistic replacement of all vehicular fuel with biofuels by using algae that have a natural oil content greater than 50%, which Briggs suggests can be grown on algae ponds at waste-water treatment plants. This oil-rich algae can then be extracted from the system and processed into biofuels, with the dried remainder further reprocessed to create ethanol. The production of algae to harvest oil for biofuels has not yet been undertaken on a commercial scale, but feasibility studies have been conducted to arrive at the above yield estimate. In addition to its projected high yield, algaculture — unlike crop-based biofuels — does not entail a decrease in food production, since it requires neither farmland nor fresh water. Many companies are pursuing algae bioreactors for various purposes, including scaling up biofuels production to commercial levels. Prof. Rodrigo E. Teixeira from the University of Alabama in Huntsville demonstrated the extraction of biofuels lipids from wet algae using a simple and economical reaction in ionic liquids.

Jatropha

Several groups in various sectors are conducting research on *Jatropha curcas*, a poisonous shrub-like tree that produces seeds considered by many to be a viable source of biofuels feedstock oil. Much of this research focuses on improving the overall per acre oil yield of Jatropha through advancements in genetics, soil science, and horticultural practices.

SG Biofuels, a San Diego-based jatropha developer, has used molecular breeding and biotechnology to produce elite hybrid seeds that show significant yield improvements over first-generation varieties. SG Biofuels also claims additional benefits have arisen from such strains, including improved flowering synchronicity, higher resistance to pests and diseases, and increased cold-weather tolerance.

Plant Research International, a department of the Wageningen University and Research Centre in the Netherlands, maintains an ongoing Jatropha Evaluation Project that examines the feasibility of large-scale jatropha cultivation through field and laboratory experiments. The Center for Sustainable Energy Farming (CfSEF) is a Los Angeles-based nonprofit research organization dedicated to jatropha research in the areas of plant science, agronomy, and horticulture. Successful exploration of these disciplines is projected to increase jatropha farm production yields by 200-300% in the next 10 years.

Fungi

A group at the Russian Academy of Sciences in Moscow, in a 2008 paper, stated they had isolated large amounts of lipids from single-celled fungi and turned it into biofuels in an economically efficient manner. More research on this fungal species, *Cunninghamella japonica*, and others, is likely to appear in the near future. The recent discovery of a variant of the fungus *Gliocladium roseum* (later renamed Ascocoryne sarcoides) points toward the production of so-called myco-diesel from cellulose. This organism was recently discovered in the rainforests of northern Patagonia, and has the unique capability of converting cellulose into medium-length hydrocarbons typically found in diesel fuel. Many other fungi that can degrade cellulose and other polymers have been observed to produce molecules that are currently being engineered using organisms from other kingdoms, suggesting that fungi may play a large role in the bio-production of fuels in the future (reviewed in).

Animal Gut Bacteria

Microbial gastrointestinal flora in a variety of animals have shown potential for the production of biofuels. Recent research has shown that TU-103, a strain of *Clostridium* bacteria found in Zebra feces, can convert nearly any form of cellulose into butanol fuel. Microbes in panda waste are being investigated for their use in creating biofuels from bamboo and other plant materials. There has also been substantial research into the technology of using the gut microbiomes of wood-feeding insects for the conversion of lignocellulotic material into biofuel.

Greenhouse Gas Emissions

Some scientists have expressed concerns about land-use change in response to greater demand for crops to use for biofuel and the subsequent carbon emissions. The payback period, that is, the time it will take biofuels to pay back the carbon debt they acquire due to land-use change, has been estimated to be between 100 and 1000 years, depending on the specific instance and location of land-use change. However, no-till practices combined with cover-crop practices can reduce the payback period to three years for grassland conversion and 14 years for forest conversion.

A study conducted in the Tocantis State, in northern Brazil, found that many families were cutting down forests in order to produce two conglomerates of oilseed plants, the J. curcas (JC group) and the R. communis (RC group). This region is composed of 15% Amazonian rainforest with high biodiversity, and 80% Cerrado forest with lower biodiversity. During the study, the farmers that planted the JC group released over 2193 Mg CO_2, while losing 53-105 Mg CO_2 sequestration from deforestation; and the RC group farmers released 562 Mg CO_2, while losing 48-90 Mg CO_2 to be sequestered from forest depletion. The production of these types of biofuels not only led into an increased emission of carbon dioxide, but also to lower efficiency of forests to absorb the gases that these farms were emitting. This has to do with the amount of fossil fuel the production of fuel crops involves. In addition, the intensive use of monocropping agriculture requires large amounts of water irrigation, as well as of fertilizers, herbicides and pesticides. This does not only lead to poisonous chemicals to disperse on water runoff, but also to the emission of nitrous oxide (NO_2) as a fertilizer byproduct, which is three hundred times more efficient in producing a greenhouse effect than carbon dioxide (CO_2).

Converting rainforests, peatlands, savannas, or grasslands to produce food crop–based biofuels in Brazil, Southeast Asia, and the United States creates a "biofuel carbon debt" by releasing 17 to 420 times more CO_2 than the annual greenhouse gas (GHG) reductions that these biofuels would provide by displacing fossil fuels. Biofuels made from waste biomass or from biomass grown on abandoned agricultural lands incur little to no carbon debt.

Water Use

In addition to water required to grow crops, biofuel facilities require significant process water.

Nuclear Power

The 1200 MWe, Leibstadt fission-electric power station in Switzerland. The boiling water reactor (BWR), located inside the dome capped cylindrical structure, is dwarfed in size by its cooling tower. The station produces a yearly average of 25 million kilowatt-hours per day, sufficient to power a city the size of Boston.

The Palo Verde Nuclear Generating Station, the largest in the US with 3 pressurized water reactors(PWRs), is situated in the Arizona desert. It uses sewage from cities as its cooling water in 9 squat mechanical draft cooling towers. Its total spent fuel/"waste" inventory produced since 1986, is contained in dry cask storage cylinders located between the artificial body of water and the electrical switchyard.

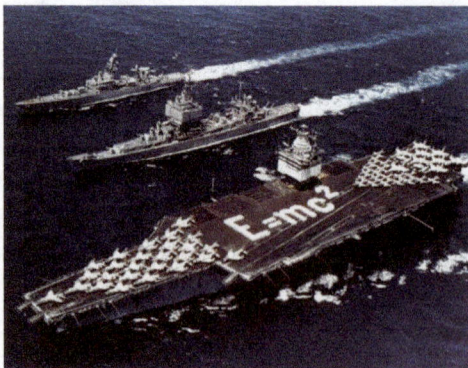

U.S. nuclear powered ships,(top to bottom) cruisers USS *Bainbridge*, the USS *Long Beach* and the *USS Enterprise*, the longest ever naval vessel, and the first nuclear-powered aircraft carrier. Picture taken in 1964 during a record setting voyage of 26,540 nmi (49,190 km) around the world in 65 days without refueling. Crew members are spelling out Einstein's mass-energy equivalence formula $E = mc^2$ on the flight deck.

The Russian nuclear-powered icebreaker NS Yamal on a joint scientific expedition with the NSF in 1994.

Nuclear power is the use of nuclear reactions that release nuclear energy to generate heat, which most frequently is then used in steam turbines to produce electricity in a nuclear power plant. The term includes nuclear fission, nuclear decay and nuclear fusion. Presently, the nuclear fission of elements in the actinide series of the periodic table produce the vast majority of nuclear energy in the direct service of humankind, with nuclear decay processes, primarily in the form of geothermal energy, and radioisotope thermoelectric generators, in niche uses making up the rest.

Nuclear fission power is a low carbon power generation method of producing electricity, and gives similar greenhouse gas emissions per unit of energy generated to renewable energy. As all electricity supplying technologies use cement etc., during construction, emissions are yet to be brought to zero. A 2014 analysis of the carbon footprint literature by the Intergovernmental Panel on Climate Change (IPCC) reported that fission electricities embodied total life-cycle emission intensity value of 12 g CO_2 eq/kWh is the lowest out of all commercial baseload energy sources, and second lowest out of all commercial electricity technologies known, after wind power which is an Intermittent energy source with embodied greenhouse gas emissions, per unit of energy generated of 11 g CO_2eq/kWh. Each result is contrasted with coal & fossil gas at 820 and 490 g CO_2 eq/kWh. With this translating into, from the beginning of Fission-electric power station commercialization in the 1970s, having prevented the emission of about 64 billion tonnes of carbon dioxide equivalent, greenhouse gases that would have otherwise resulted from the burning of fossil fuels in thermal power stations.

There is a social debate about nuclear power. Proponents, such as the World Nuclear Association and Environmentalists for Nuclear Energy, contend that nuclear power is a safe, sustainable energy source that reduces carbon emissions. Opponents, such as Greenpeace International and NIRS, contend that nuclear power poses many threats to people and the environment.

Far-reaching fission power reactor accidents, or accidents that resulted in medium to long-lived fission product contamination of inhabited areas, have occurred in Generation I & II reactor designs, blueprinted between 1950 and 1980. These include the Chernobyl disaster which occurred in 1986, the Fukushima Daiichi nuclear disaster (2011), and the more contained Three Mile Island accident (1979). There have also been some nuclear submarine accidents. In terms of lives lost per unit of energy generated, analysis has determined that fission-electric reactors have caused fewer fatalities per unit of energy generated than the other major sources of energy generation. Energy production from coal, petroleum, natural gas and hydroelectricity has caused a greater number of fatalities per unit of energy generated due to air pollution and energy accident effects. Four years after the Fukushima-Daiichi accident, there have been no fatalities due to exposure to radiation, and no discernible increased incidence of radiation-related health effects are expected among exposed members of the public and their descendants. The Japan Times estimated 1,600 deaths were the result of evacuation, due to physical and mental stress stemming from long stays at shelters, a lack of initial care as a result of hospitals being disabled by the disaster, and suicides.

In 2015:

- Ten new reactors were connected to the grid.

- Seven reactors were permanently shut down.

- 441 reactors had a worldwide net capacity of 382,855 megawatts of electricity.

- 67 new nuclear reactors were under construction.

Most of the new activity is in China where there is an urgent need to control pollution from coal plants.

History

Origins

In 1932 physicist Ernest Rutherford discovered that when lithium atoms were "split" by protons from a proton accelerator, immense amounts of energy were released in accordance with the principle of mass–energy equivalence. However, he and other nuclear physics pioneers Niels Bohr and Albert Einstein believed harnessing the power of the atom for practical purposes anytime in the near future was unlikely, with Rutherford labeling such expectations "moonshine."

The same year, his doctoral student James Chadwick discovered the neutron, which was immediately recognized as a potential tool for nuclear experimentation because of its lack of an electric charge. Experimentation with bombardment of materials with neutrons led Frédéric and Irène Joliot-Curie to discover induced radioactivity in 1934, which allowed the creation of radium-like elements at much less the price of natural radium. Further work by Enrico Fermi in the 1930s focused on using slow neutrons to increase the effectiveness of induced radioactivity. Experiments bombarding uranium with neutrons led Fermi to believe he had created a new, transuranic element, which was dubbed hesperium.

December 2, 1942. A depiction of the scene when scientists observed the world's first man made nuclear reactor, the Chicago Pile-1, as it became self-sustaining/critical at the University of Chicago.

But in 1938, German chemists Otto Hahn and Fritz Strassmann, along with Austrian physicist Lise Meitner and Meitner's nephew, Otto Robert Frisch, conducted experiments with the products of neutron-bombarded uranium, as a means of further investigating Fermi's claims. They determined that the relatively tiny neutron split the nucleus of the massive uranium atoms into two roughly equal pieces, contradicting Fermi. This was an extremely surprising result: all other forms of nuclear decay involved only small changes to the mass of the nucleus, whereas this process—dubbed "fission" as a reference to biology—involved a complete rupture of the nucleus. Numerous scientists, including Leó Szilárd, who was one of the first, recognized that if fission reactions released additional neutrons, a self-sustaining nuclear chain reaction could result. Once this was experimentally confirmed and announced by Frédéric Joliot-Curie in 1939, scientists in many countries (including the United States, the United Kingdom, France, Germany, and the Soviet Union) petitioned their governments for support of nuclear fission research, just on the cusp of World War II, for the development of a nuclear weapon.

In the United States, where Fermi and Szilárd had both emigrated, this led to the creation of the first man-made reactor, known as Chicago Pile-1, which achieved criticality on December 2, 1942. This work became part of the Manhattan Project, which made enriched uranium and built large reactors to breed plutonium for use in the first nuclear weapons, which were used on the cities of Hiroshima and Nagasaki.

The first light bulbs ever lit by electricity generated by nuclear power at EBR-1 at Argonne
National Laboratory-West, December 20, 1951.

In 1945, the pocketbook *The Atomic Age* heralded the untapped atomic power in everyday objects and depicted a future where fossil fuels would go unused. One science writer, David Dietz, wrote that instead of filling the gas tank of your car two or three times a week, you will travel for a year on a pellet of atomic energy the size of a vitamin pill. Glenn Seaborg, who chaired the Atomic Energy Commission, wrote "there will be nuclear powered earth-to-moon shuttles, nuclear powered artificial hearts, plutonium heated swimming pools for SCUBA divers, and much more". These overly optimistic predications remain unfulfilled.

United Kingdom, Canada, and USSR proceeded over the course of the late 1940s and early 1950s. Electricity was generated for the first time by a nuclear reactor on December 20, 1951, at the EBR-I experimental station near Arco, Idaho, which initially produced about 100 kW. Work was also strongly researched in the US on nuclear marine propulsion, with a test reactor being developed by 1953 (eventually, the USS Nautilus, the first nuclear-powered submarine, would launch in 1955). In 1953, US President Dwight Eisenhower gave his "Atoms for Peace" speech at the United Nations, emphasizing the need to develop "peaceful" uses of nuclear power quickly. This was followed by the 1954 Amendments to the Atomic Energy Act which allowed rapid declassification of U.S. reactor technology and encouraged development by the private sector.

Early Years

On June 27, 1954, the USSR's Obninsk Nuclear Power Plant became the world's first nuclear power plant to generate electricity for a power grid, and produced around 5 megawatts of electric power.

Later in 1954, Lewis Strauss, then chairman of the United States Atomic Energy Commission (U.S. AEC, forerunner of the U.S. Nuclear Regulatory Commission and the United States Department of Energy) spoke of electricity in the future being "too cheap to meter". Strauss was very likely referring to hydrogen fusion —which was secretly being developed as part of Project Sherwood at the time—but Strauss's statement was interpreted as a promise of very cheap energy from nuclear fission. The U.S. AEC itself had issued far more realistic testimony regarding nuclear fission to the U.S. Congress only months before, projecting that "costs can be brought down... [to]... about the same as the cost of electricity from conventional sources..."

In 1955 the United Nations' "First Geneva Conference", then the world's largest gathering of scientists and engineers, met to explore the technology. In 1957 EURATOM was launched alongside the European Economic Community (the latter is now the European Union). The same year also saw the launch of the International Atomic Energy Agency (IAEA).

Calder Hall, United Kingdom - The world's first commercial nuclear power station. First connected to the national power grid on 27 August 1956 and officially opened by Queen Elizabeth II on 17 October 1956

The Shippingport Atomic Power Station in Shippingport, Pennsylvania was the first commercial reactor in the USA and was opened in 1957.

The world's first commercial nuclear power station, Calder Hall at Windscale, England, was opened in 1956 with an initial capacity of 50 MW (later 200 MW). The first commercial nuclear generator to become operational in the United States was the Shippingport Reactor (Pennsylvania, December 1957).

One of the first organizations to develop nuclear power was the U.S. Navy, for the purpose of propelling submarines and aircraft carriers. The first nuclear-powered submarine, USS *Nautilus* (SSN-571), was put to sea in December 1954. As of 2016, the U.S. Navy submarine fleet is made up entirely of nuclear-powered vessels, with 75 submarines in service. Two U.S. nuclear submarines, USS *Scorpion* and USS *Thresher*, have been lost at sea. The Russian Navy is currently (2016) estimated to have 61 nuclear submarines in service; eight Soviet and Russian nuclear submarines have been lost at sea. This includes the Soviet submarine K-19 reactor accident in 1961 which resulted in 8 deaths and more than 30 other people were over-exposed to radiation. The Soviet submarine K-27 reactor accident in 1968 resulted in 9 fatalities and 83 other injuries. Moreover, Soviet submarine K-429 sank twice, but was raised after each incident. Several serious nuclear and radiation accidents have involved nuclear submarine mishaps.

The U.S. Army also had a nuclear power program, beginning in 1954. The SM-1 Nuclear Power Plant, at Fort Belvoir, Virginia, was the first power reactor in the U.S. to supply electrical energy to a commercial grid (VEPCO), in April 1957, before Shippingport. The SL-1 was a U.S. Army experimental nuclear power reactor at the National Reactor Testing Station in eastern Idaho. It underwent a steam explosion and meltdown in January 1961, which killed its three operators. In Soviet Union in The Mayak Production Association there were a number of accidents including an explosion that released 50-100 tonnes of high-level radioactive waste, contaminating a huge territory in the eastern Urals and causing numerous deaths and injuries. The Soviet regime kept this accident secret for about 30 years. The event was eventually rated at 6 on the seven-level INES scale (third in severity only to the disasters at Chernobyl and Fukushima).

Development

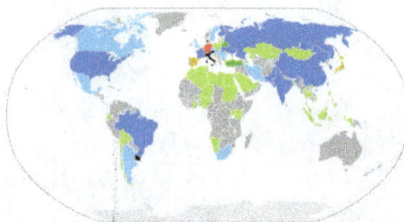

The status of nuclear power globally (click image for legend)

Washington Public Power Supply System Nuclear Power Plants 3 and 5 were never completed.

Installed nuclear capacity initially rose relatively quickly, rising from less than 1 gigawatt (GW) in 1960 to 100 GW in the late 1970s, and 300 GW in the late 1980s. Since the late 1980s worldwide capacity has risen much more slowly, reaching 366 GW in 2005. Between around 1970 and 1990, more than 50 GW of capacity was under construction (peaking at over 150 GW in the late 1970s and early 1980s) — in 2005, around 25 GW of new capacity was planned. More than two-thirds of all nuclear plants ordered after January 1970 were eventually cancelled. A total of 63 nuclear units were canceled in the USA between 1975 and 1980.

During the 1970s and 1980s rising economic costs (related to extended construction times largely due to regulatory changes and pressure-group litigation) and falling fossil fuel prices made nuclear power plants then under construction less attractive. In the 1980s (U.S.) and 1990s (Europe), flat load growth and electricity liberalization also made the addition of large new baseload capacity unattractive.

The 1973 oil crisis had a significant effect on countries, such as France and Japan, which had relied more heavily on oil for electric generation (39% and 73% respectively) to invest in nuclear power.

Some local opposition to nuclear power emerged in the early 1960s, and in the late 1960s some members of the scientific community began to express their concerns. These concerns related to nuclear accidents, nuclear proliferation, high cost of nuclear power plants, nuclear terrorism and radioactive waste disposal. In the early 1970s, there were large protests about a proposed nuclear power plant in Wyhl, Germany. The project was cancelled in 1975 and anti-nuclear success at Wyhl inspired opposition to nuclear power in other parts of Europe and North America. By the mid-

1970s anti-nuclear activism had moved beyond local protests and politics to gain a wider appeal and influence, and nuclear power became an issue of major public protest. Although it lacked a single co-ordinating organization, and did not have uniform goals, the movement's efforts gained a great deal of attention. In some countries, the nuclear power conflict "reached an intensity unprecedented in the history of technology controversies".

120,000 people attended an anti-nuclear protest in Bonn, Germany, on October 14, 1979, following the Three Mile Island accident.

In France, between 1975 and 1977, some 175,000 people protested against nuclear power in ten demonstrations. In West Germany, between February 1975 and April 1979, some 280,000 people were involved in seven demonstrations at nuclear sites. Several site occupations were also attempted. In the aftermath of the Three Mile Island accident in 1979, some 120,000 people attended a demonstration against nuclear power in Bonn. In May 1979, an estimated 70,000 people, including then governor of California Jerry Brown, attended a march and rally against nuclear power in Washington, D.C. Anti-nuclear power groups emerged in every country that has had a nuclear power programme.

Three Mile Island and Chernobyl

The abandoned city of Pripyat with Chernobyl plant in the distance.

Health and safety concerns, the 1979 accident at Three Mile Island, and the 1986 Chernobyl disaster played a part in stopping new plant construction in many countries, although the public policy organization, the Brookings Institution states that new nuclear units, at the time of publishing in 2006, had not been built in the U.S. because of soft demand for electricity, and cost overruns on nuclear plants due to regulatory issues and construction delays. By the end of the 1970s it became clear that nuclear power would not grow nearly as dramatically as once believed. Eventually, more than 120 reactor orders in the U.S. were ultimately cancelled and the construction of new reactors ground to a halt. A cover story in the February 11, 1985, issue of *Forbes* magazine commented on the overall failure of the U.S. nuclear power program, saying it "ranks as the largest managerial disaster in business history".

Unlike the Three Mile Island accident, the much more serious Chernobyl accident did not increase regulations affecting Western reactors since the Chernobyl reactors were of the problematic RBMK design only used in the Soviet Union, for example lacking "robust" containment buildings. Many of these RBMK reactors are still in use today. However, changes were made in both the reactors themselves (use of a safer enrichment of uranium) and in the control system (prevention of disabling safety systems), amongst other things, to reduce the possibility of a duplicate accident.

An international organization to promote safety awareness and professional development on operators in nuclear facilities was created: WANO; World Association of Nuclear Operators.

Opposition in Ireland and Poland prevented nuclear programs there, while Austria (1978), Sweden (1980) and Italy (1987) (influenced by Chernobyl) voted in referendums to oppose or phase out nuclear power. In July 2009, the Italian Parliament passed a law that cancelled the results of an earlier referendum and allowed the immediate start of the Italian nuclear program. After the Fukushima Daiichi nuclear disaster a one-year moratorium was placed on nuclear power development, followed by a referendum in which over 94% of voters (turnout 57%) rejected plans for new nuclear power.

Nuclear Renaissance

Olkiluoto 3 under construction in 2009. It is the first EPR design, but problems with workmanship and supervision have created costly delays which led to an inquiry by the Finnish nuclear regulator STUK. In December 2012, Areva estimated that the full cost of building the reactor will be about €8.5 billion, or almost three times the original delivery price of €3 billion.

Since about 2001 the term *nuclear renaissance* has been used to refer to a possible nuclear power industry revival, driven by rising fossil fuel prices and new concerns about meeting greenhouse gas emission limits. In 2012, the World Nuclear Association reported that nuclear electricity generation was at its lowest level since 1999. As of January 2016, however, 65 new nuclear power reactors were under construction. Over 150 were planned, equivalent to nearly half of capacity at that time.

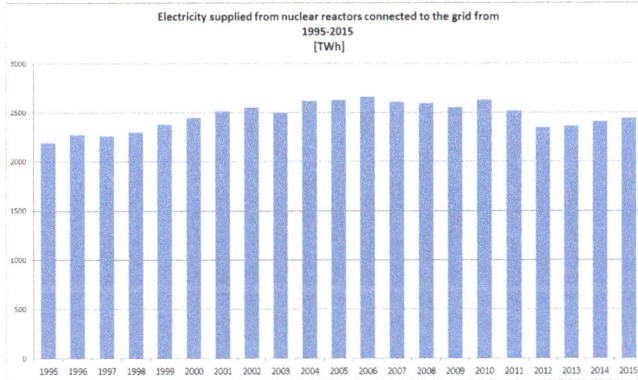

Production of nuclear power plants

Fukushima Daiichi Nuclear Disaster

Japan's 2011 Fukushima Daiichi nuclear accident, which occurred in a reactor design from the 1960s, prompted a re-examination of nuclear safety and nuclear energy policy in many countries. Germany plans to close all its reactors by 2022, and Italy has re-affirmed its ban on electric utilities generating, but not importing, fission derived electricity. In 2011 the International Energy Agency halved its prior estimate of new generating capacity to be built by 2035. In 2013 Japan signed a deal worth $22 billion, in which Mitsubishi Heavy Industries would build four modern *Atmea* reactors for Turkey. In August 2015, following 4 years of near zero fission-electricity generation, Japan began restarting its fission fleet, after safety upgrades were completed, beginning with Sendai fission-electric station.

In March 2011 the nuclear emergencies at Japan's Fukushima Daiichi Nuclear Power Plant and shutdowns at other nuclear facilities raised questions among some commentators over the future of the renaissance. China, Germany, Switzerland, Israel, Malaysia, Thailand, United Kingdom, Italy and the Philippines have reviewed their nuclear power programs. Indonesia and Vietnam still plan to build nuclear power plants.

The World Nuclear Association has said that "nuclear power generation suffered its biggest ever one-year fall through 2012 as the bulk of the Japanese fleet remained offline for a full calendar year". Data from the International Atomic Energy Agency showed that nuclear power plants globally produced 2346 TWh of electricity in 2012 – seven per cent less than in 2011. The figures illustrate the effects of a full year of 48 Japanese power reactors producing no power during the year. The permanent closure of eight reactor units in Germany was also a factor. Problems at Crystal River, Fort Calhoun and the two San Onofre units in the USA meant they produced no power for the full year, while in Belgium Doel 3 and Tihange 2 were out of action for six months. Compared to 2010, the nuclear industry produced 11% less electricity in 2012.

Post-Fukushima Controversy

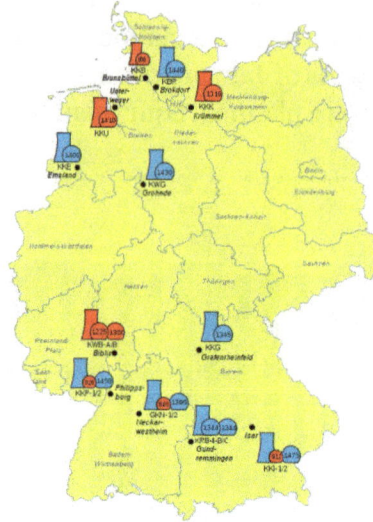

Eight of the seventeen operating reactors in Germany were permanently shut down following
the March 2011 Fukushima nuclear disaster.

The Fukushima Daiichi nuclear accident sparked controversy about the import of the accident and its effect on nuclear's future. IAEA Director General Yukiya Amano said the Japanese nuclear accident "caused deep public anxiety throughout the world and damaged confidence in nuclear power", and the International Energy Agency halved its estimate of additional nuclear generating capacity to be built by 2035. But by 2015, the Agency's outlook had become more promising. "Nuclear power is a critical element in limiting greenhouse gas emissions," the agency noted, and "the prospects for nuclear energy remain positive in the medium to long term despite a negative impact in some countries in the aftermath of the [Fukushima-Daiichi] accident...it is still the second-largest source worldwide of low-carbon electricity. And the 72 reactors under construction at the start of last year were the most in 25 years." Though Platts reported in 2011 that "the crisis at Japan's Fukushima nuclear plants has prompted leading energy-consuming countries to review the safety of their existing reactors and cast doubt on the speed and scale of planned expansions around the world", Progress Energy Chairman/CEO Bill Johnson made the observation that "Today there's an even more compelling case that greater use of nuclear power is a vital part of a balanced energy strategy". In 2011, *The Economist* opined that nuclear power "looks dangerous, unpopular, expensive and risky", and that "it is replaceable with relative ease and could be forgone with no huge structural shifts in the way the world works". Earth Institute Director Jeffrey Sachs disagreed, claiming combating climate change would require an expansion of nuclear power. "We won't meet the carbon targets if nuclear is taken off the table," he said. "We need to understand the scale of the challenge."

Investment banks were critical of nuclear soon after the accident. Many disputed their impartiality, however, due to significant investments in renewable energy, perceived by some as a valid alternative to nuclear. In early April 2011, analysts at Swiss-based investment bank UBS said: "At Fukushima, four reactors have been out of control for weeks, casting doubt on whether even an advanced economy can master nuclear safety...we believe the Fukushima accident was the most serious ever for the credibility of nuclear power". UBS has helped to raise more than $20 billion since 2006 and advised on more than a dozen deals for renewable energy and cleantech companies. Deutsche Bank advised that "the global impact of the Fukushima accident is a fundamental shift in public perception with regard to how a nation prioritizes and values its populations health, safety, security, and natural environment when determining its current and future energy pathways...renewable energy will be a clear long-term winner in most energy systems, a conclusion supported by many voter surveys conducted over the past few weeks. Deutsche Bank has over €1 billion in capital invested in renewables projects in Europe, North & South America, and Asia.

Manufacturers also recognized a profit opportunity in negative public perceptions about nuclear. In September 2011, German engineering giant Siemens announced it will withdraw entirely from the nuclear industry, as a response to the Fukushima nuclear accident in Japan, and said that it would no longer build nuclear power plants anywhere in the world. The company's chairman, Peter Löscher, said that "Siemens was ending plans to cooperate with Rosatom, the Russian state-controlled nuclear power company, in the construction of dozens of nuclear plants throughout Russia over the coming two decades". Renewable energy is a core component of Siemens's profit base. In February, 2016 the firm proposed a €10 billion renewable energy investment in Egypt.

In February 2012, the United States Nuclear Regulatory Commission approved the construction of two additional reactors at the Vogtle Electric Generating Plant, the first reactors to be approved in over 30 years since the Three Mile Island accident, but NRC Chairman Gregory Jaczko cast a dissenting vote citing safety concerns stemming from Japan's 2011 Fukushima nuclear disaster, and saying "I cannot support issuing this license as if Fukushima never happened". Jaczko resigned in April 2012. One week after Southern received the license to begin major construction on the two new reactors, a dozen environmental and anti-nuclear groups sued to stop the Plant Vogtle expansion project, saying "public safety and environmental problems since Japan's Fukushima Daiichi nuclear reactor accident have not been taken into account". In July 2012, the suit was rejected by the Washington, D.C. Circuit Court of Appeals.

Countries such as Australia, Austria, Denmark, Greece, Ireland, Italy, Latvia, Liechtenstein, Luxembourg, Malta, Portugal, Israel, Malaysia, New Zealand, and Norway have no nuclear power reactors and remain opposed to nuclear power. However, by contrast, some countries remain in favor, and financially support nuclear fusion research, including EU wide funding of the ITER project.

Industry

Capacity and Production

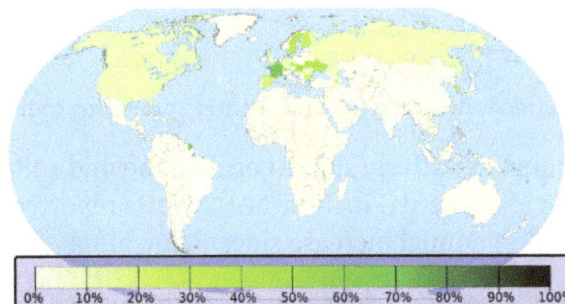

Percentage of a nations electricity, produced by fission-electric power stations.

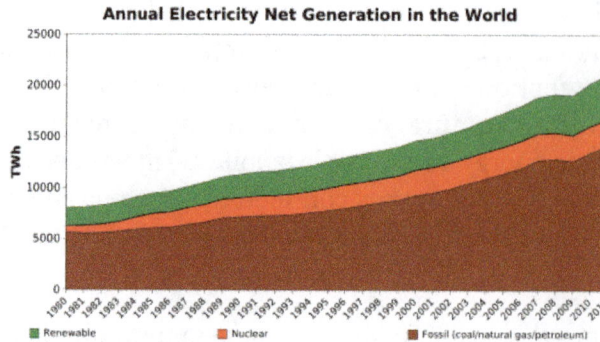

Annual Electricity Net Generation in the World

Net electrical generation by source and growth from 1980 to 2010. (Brown) - fossil fuels.(Red) - Fission.(Green)- "all renewables". In terms of energy generated between 1980 and 2010, the contribution from fission grew the fastest.

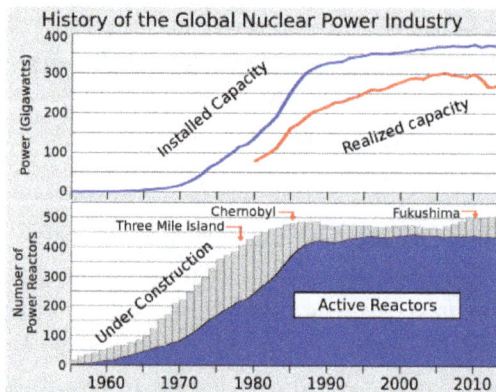

History of the Global Nuclear Power Industry

The rate of new construction builds for civilian fission-electric reactors essentially halted in the late 1980s, with the effects of accidents having a chilling effect. Increased capacity factor realizations in existing reactors was primarily responsible for the continuing increase in electrical energy produced during this period. The halting of new builds c. 1985, resulted in greater fossil fuel generation.

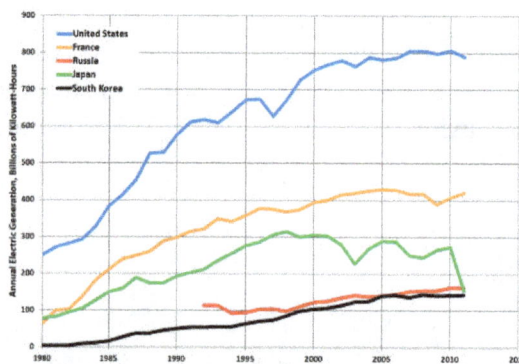

Electricity generation trends in the top five fission-energy producing countries (US EIA data)

Nuclear power capacity remained relatively stable between the mid 1980s until the accident at the Fukushima Daiichi reactor in March 2011. In June 2015, Platts reported global nuclear generation increased by 1% in 2014, the first annual increase since Fukushima.

The United States produces the most nuclear energy, with nuclear power providing 19% of the electricity it consumes, while France produces the highest percentage of its electrical energy from nuclear reactors—80% as of 2006. In the European Union as a whole, nuclear energy provides 30% of the electricity. Nuclear energy policy differs among European Union countries, and some, such as Austria, Estonia, Ireland and Italy, have no active nuclear power stations. In comparison, France has a large number of these plants, with 16 multi-unit stations in current use.

Many military and some civilian (such as some icebreaker) ships use nuclear marine propulsion, a form of nuclear propulsion. A few space vehicles have been launched using full-fledged nuclear reactors: 33 reactors belong to the Soviet RORSAT series and one was the American SNAP-10A.

International research is continuing into safety improvements such as passively safe plants, the use of nuclear fusion, and additional uses of process heat such as hydrogen production (in support of a hydrogen economy), for desalinating sea water, and for use in district heating systems.

Nuclear (fission) power stations, excluding the contribution from naval nuclear fission reactors, provided 11% of the world's electricity in 2012, somewhat less than that generated by hydro-electric stations at 16%. Since electricity accounts for about 25% of humanity's energy usage with the majority of the rest coming from fossil fuel reliant sectors such as transport, manufacture and home heating, nuclear fission's contribution to the global final energy consumption is about 2.5%, a little more than the combined global electricity production from "new renewables"; wind, solar, biofuel and geothermal power, which together provided 2% of global final energy consumption in 2014.

Regional differences in the use of fission energy are large. Fission energy generation, with a 20% share of the U.S. electricity production, is the single largest deployed technology among current low-carbon power sources in the country. In addition, two-thirds of the European Union's twenty-seven nations' low-carbon energy is produced by fission. Some of these nations have banned its generation, such as Italy, which ended the use of fission-electric generation, which started in 1963, in 1990. France is the largest user of nuclear energy, deriving 75% of its electricity from fission.

In 2013, the IAEA reported that there were 437 operational civil fission-electric reactors in 31 countries, although not every reactor was producing electricity. In addition, there were approximately 140 naval vessels using nuclear propulsion in operation, powered by some 180 reactors. As of 2013, attaining a net energy gain from sustained nuclear fusion reactions, excluding natural fusion power sources such as the Sun, remains an ongoing area of international physics and engineering research. With commercial fusion power production remaining unlikely before 2050.

Since commercial nuclear energy began in the mid-1950s, 2008 was the first year that no new nuclear power plant was connected to the grid, although two were connected in 2009.

In 2015, the IAEA reported that worldwide there were 67 civil fission-electric power reactors under construction in 15 countries including Gulf states such as the United Arab Emirates (UAE). Over half of the 67 total being built were in Asia, with 28 in China. Eight new grid connections were completed by China in 2015 and the most recently completed reactor to be connected to the electrical grid, as of January 2016, was at the Kori Nuclear Power Plant in the Republic of Korea. In the US, four new Generation III reactors were under construction at Vogtle and Summer station, while a fifth was nearing completion at Watts Bar station, all five were expected to become operational before 2020. In 2013, four aging uncompetitive U.S reactors were closed. According to the World Nuclear Association, the global trend is for new nuclear power stations coming online to be balanced by the number of old plants being retired.

Economics

George W. Bush signing the Energy Policy Act of 2005, which was designed to promote the US nu-clear power industry, through incentives and subsidies, including cost-overrun support up to a to-tal of $2 billion for six new nuclear plants. However, as of 2014 some electric utilities have rebuffed the loan package, including South Carolina Electric and Gas which operates Summer Station (the location of 2 new builds), noting instead that "it was easier to raise [loan] money commercially."

The Ikata Nuclear Power Plant, a pressurized water reactor that cools by utilizing a secondary coolant heat exchanger with a large body of water, an alternative cooling approach to large cooling towers.

Nuclear power plants typically have high capital costs for building the plant, but low fuel costs. Although nuclear power plants can vary their output the electricity is generally less favorably priced when doing so. Nuclear power plants are therefore typically run as much as possible to keep the cost of the generated electrical energy as low as possible, supplying mostly base-load electricity.

Internationally the price of nuclear plants rose 15% annually in 1970-1990. Yet, nuclear power has total costs in 2012 of about $96 per megawatt hour (MWh), most of which involves capital construction costs, compared with solar power at $130 per MWh, and natural gas at the low end at $64 per MWh.

In 2015, the *Bulletin of the Atomic Scientists* unveiled the Nuclear Fuel Cycle Cost Calculator, an online tool that estimates the full cost of electricity produced by three configurations of the nuclear fuel cycle. Two years in the making, this interactive calculator is the first generally accessible model to provide a nuanced look at the economic costs of nuclear power; it lets users test how sensitive the price of electricity is to a full range of components—more than 60 parameters that can be adjusted for the three configurations of the nuclear fuel cycle considered by this tool (once-through, limited-recycle, full-recycle). Users can select the fuel cycle they would like to examine, change cost estimates for each component of that cycle, and even choose uncertainty ranges for the cost of particular components. This approach allows users around the world to compare the cost of different nuclear power approaches in a sophisticated way, while taking account of prices relevant to their own countries or regions.

In recent years there has been a slowdown of electricity demand growth. In Eastern Europe, a number of long-established projects are struggling to find finance, notably Belene in Bulgaria and the additional reactors at Cernavoda in Romania, and some potential backers have pulled out. Where the electricity market is competitive, cheap natural gas is available, and its future supply relatively secure, this also poses a major problem for nuclear projects and existing plants.

Analysis of the economics of nuclear power must take into account who bears the risks of future uncertainties. To date all operating nuclear power plants were developed by state-owned or regulated utility monopolies where many of the risks associated with construction costs, operating performance, fuel price, accident liability and other factors were borne by consumers rather than suppliers. In addition, because the potential liability from a nuclear accident is so great, the full cost of liability insurance is generally limited/capped by the government, which the U.S. Nuclear Regulatory Commission concluded constituted a significant subsidy. Many countries have now liberalized the electricity market where these risks, and the risk of cheaper competitors emerging before capital costs are recovered, are borne by plant suppliers and operators rather than consumers, which leads to a significantly different evaluation of the economics of new nuclear power plants.

Following the 2011 Fukushima Daiichi nuclear disaster, costs are expected to increase for currently operating and new nuclear power plants, due to increased requirements for on-site spent fuel management and elevated design basis threats.

The economics of new nuclear power plants is a controversial subject, since there are diverging views on this topic, and multibillion-dollar investments ride on the choice of an energy source. Comparison with other power generation methods is strongly dependent on assumptions about construction timescales and capital financing for nuclear plants as well as the future costs of fossil fuels and renewables as well as for energy storage solutions for intermittent power sources. Cost estimates also need to take into account plant decommissioning and nuclear waste storage costs. On the other hand, measures to mitigate global warming, such as a carbon tax or carbon emissions trading, may favor the economics of nuclear power.

Nuclear Power Organizations

There are multiple organizations which have taken a position on nuclear power and the nuclear power industry– some are proponents, and some are opponents.

Proponents

The majority of pro-nuclear energy organizations and associations is either industry-supported or directly formed from industry members as advocacy groups or trade associations.

- Environmentalists for Nuclear Energy (International)

- Nuclear Industry Association (United Kingdom)

- World Nuclear Association, a confederation of companies connected with nuclear power production. (International)

- International Atomic Energy Agency (IAEA)

- Nuclear Energy Institute (United States)

- American Nuclear Society (United States)

- United Kingdom Atomic Energy Authority (United Kingdom)

- EURATOM (Europe)

- European Nuclear Education Network (Europe)

- Atomic Energy of Canada Limited (Canada)

- Nuclear Matters (United States)

- Breakthrough Institute (United States)

- Thorium Energy Alliance (United States)

- Californians for Green Nuclear Power (United States)

- Save Diablo Canyon (United States)

- Thorium Now (United States)

- Category:Nuclear industry organizations

Opponents

- Friends of the Earth International, a network of environmental organizations.

- Greenpeace International, a non-governmental organization

- Nuclear Information and Resource Service (International)

- World Information Service on Energy (International)

- Sortir du nucléaire (France)

- Pembina Institute (Canada)

- Institute for Energy and Environmental Research (United States)

- Sayonara Nuclear Power Plants (Japan)

- Category:Anti-nuclear organizations

Future of the Industry

Brunswick Nuclear Plant discharge canal

The Bruce Nuclear Generating Station, the largest nuclear power facility in the world

The nuclear power industry in western nations has a history of construction delays, cost overruns, plant cancellations, and nuclear safety issues despite significant government subsidies and support. In December 2013, *Forbes* magazine reported that, in developed countries, "reactors are not a viable source of new power". Even in developed nations where they make economic sense, they are not feasible because nuclear's "enormous costs, political and popular opposition, and regulatory uncertainty". This view echoes the statement of former Exelon CEO John Rowe, who said in 2012 that new nuclear plants "don't make any sense right now" and won't be economically viable in the foreseeable future. John Quiggin, economics professor, also says the main problem with the nuclear option is that it is not economically-viable. Quiggin says that we need more efficient energy use and more renewable energy commercialization. Former NRC member Peter Bradford and Professor Ian Lowe have recently made similar statements. However, some "nuclear cheerleaders" and lobbyists in the West continue to champion reactors, often with proposed new but largely untested designs, as a source of new power.

Much more new build activity is occurring in developing countries like South Korea, India and China. In March 2016, China had 30 reactors in operation, 24 under construction and plans to build more, However, according to a government research unit, China must not build "too many nuclear power reactors too quickly", in order to avoid a shortfall of fuel, equipment and qualified plant workers.

In the US, licenses of almost half its reactors have been extended to 60 years, Two new Generation III reactors are under construction at Vogtle, a dual construction project which marks the end of a 34-year period of stagnation in the US construction of civil nuclear power reactors. The station operator licenses of almost half the present 104 power reactors in the US, as of 2008, have been given extensions to 60 years. As of 2012, U.S. nuclear industry officials expect five new reactors to enter service by 2020, all at existing plants. In 2013, four aging, uncompetitive, reactors were permanently closed. Relevant state legislatures are trying to close Vermont Yankee and Indian Point Nuclear Power Plant.

The U.S. NRC and the U.S. Department of Energy have initiated research into Light water reactor sustainability which is hoped will lead to allowing extensions of reactor licenses beyond 60 years, provided that safety can be maintained, as the loss in non-CO_2-emitting generation capacity by retiring reactors "may serve to challenge U.S. energy security, potentially resulting in increased greenhouse gas emissions, and contributing to an imbalance between electric supply and demand."

There is a possible impediment to production of nuclear power plants as only a few companies worldwide have the capacity to forge single-piece reactor pressure vessels, which are necessary in the most common reactor designs. Utilities across the world are submitting orders years in advance of any actual need for these vessels. Other manufacturers are examining various options, including making the component themselves, or finding ways to make a similar item using alternate methods.

According to the World Nuclear Association, globally during the 1980s one new nuclear reactor started up every 17 days on average, and in the year 2015 it was estimated that this rate could in theory eventually increase to one every 5 days, although no plans exist for that. As of 2007, Watts Bar 1 in Tennessee, which came on-line on February 7, 1996, was the last U.S. commercial nuclear reactor to go on-line. This is often quoted as evidence of a successful worldwide campaign for nuclear power phase-out. Electricity shortages, fossil fuel price increases, global warming, and heavy metal emissions from fossil fuel use, new technology such as passively safe plants, and national energy security may renew the demand for nuclear power plants.

Nuclear Power Plant

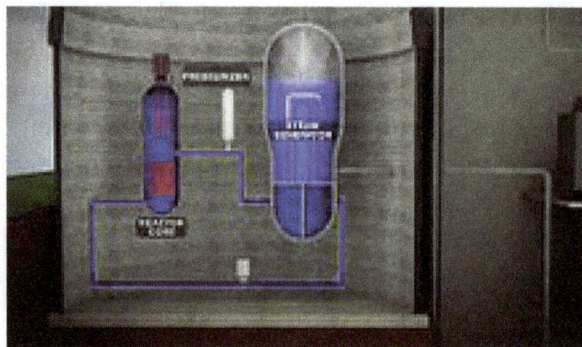

Play media

An animation of a Pressurized water reactor in operation.

Unlike fossil fuel power plants, the only substance leaving the cooling towers of nuclear power plants is water vapour and thus does not pollute the air or cause global warming.

Just as many conventional thermal power stations generate electricity by harnessing the thermal energy released from burning fossil fuels, nuclear power plants convert the energy released from the nucleus of an atom via nuclear fission that takes place in a nuclear reactor. The heat is removed from the reactor core by a cooling system that uses the heat to generate steam, which drives a steam turbine connected to a generator producing electricity.

Life Cycle of Nuclear Fuel

The nuclear fuel cycle begins when uranium is mined, enriched, and manufactured into nuclear fuel, (1) which is delivered to a nuclear power plant. After usage in the power plant, the spent fuel is delivered to a reprocessing plant (2) or to a final repository (3) for geological disposition. In reprocessing 95% of spent fuel can potentially be recycled to be returned to usage in a power plant (4).

A nuclear reactor is only part of the life-cycle for nuclear power. The process starts with mining. Uranium mines are underground, open-pit, or in-situ leach mines. In any case, the uranium ore is extracted, usually converted into a stable and compact form such as yellowcake, and then transported to a processing facility. Here, the yellowcake is converted to uranium hexafluoride, which is then enriched using various techniques. At this point, the enriched uranium, containing more than the natural 0.7% U-235, is used to make rods of the proper composition and geometry for the particular reactor that the fuel is destined for. The fuel rods will spend about 3 operational cycles (typically 6 years total now) inside the reactor, generally until about 3% of their uranium has been fissioned, then they will be moved to a spent fuel pool where the short lived isotopes generated by fission can decay away. After about 5 years in a spent fuel pool the spent fuel is radioactively and thermally cool enough to handle, and it can be moved to dry storage casks or reprocessed.

Conventional Fuel Resources

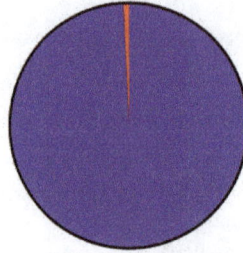

Natural uranium
> 99.2% U-238
0.72% U-235

Low-enriched uranium
(reactor grade)
3-4% U-235

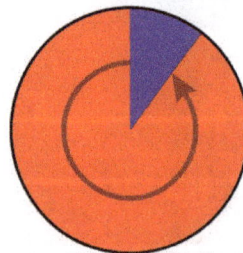

Highly enriched uranium
(weapons grade)
90% U-235

Proportions of the isotopes, uranium-238 (blue) and uranium-235 (red) found naturally,
versus grades that are enriched. light water reactors require fuel enriched to (3-4%),
while others such as the CANDU reactor uses natural uranium.

Uranium is a fairly common element in the Earth's crust. Uranium is approximately as common as tin or germanium in the Earth's crust, and is about 40 times more common than silver. Uranium is a constituent of most rocks, dirt, and of the oceans. The fact that uranium is so spread out is a problem because mining uranium is only economically feasible where there is a large concentration. Still, the world's present measured resources of uranium, economically recoverable at a price of 130 USD/kg, are enough to last for between 70 and 100 years.

According to the OECD in 2006, there is an expected 85 years worth of uranium in identified resources, when that uranium is used in present reactor technology, with 670 years of economically recoverable uranium in total conventional resources and phosphate ores, while also using present reactor technology, a resource that is recoverable from between 60-100 US$/kg of Uranium. The OECD have noted that:

Even if the nuclear industry expands significantly, sufficient fuel is available for centuries. If advanced breeder reactors could be designed in the future to efficiently utilize recycled or depleted uranium and all actinides, then the resource utilization efficiency would be further improved by an additional factor of eight.

For example, the OECD have determined that with a pure fast reactor fuel cycle with a burn up of, and recycling of, all the Uranium and actinides, actinides which presently make up the most hazardous substances in nuclear waste, there is 160,000 years worth of Uranium in total conventional resources and phosphate ore. According to the OECD's red book in 2011, due to increased exploration, known uranium resources have grown by 12.5% since 2008, with this increase translating into greater than a century of uranium available if the metals usage rate were to continue at the 2011 level.

Current light water reactors make relatively inefficient use of nuclear fuel, fissioning only the very rare uranium-235 isotope. Nuclear reprocessing can make this waste reusable, and more efficient reactor designs, such as the currently under construction Generation III reactors achieve a higher efficiency burn up of the available resources, than the current vintage generation II reactors, which make up the vast majority of reactors worldwide.

Breeding

As opposed to current light water reactors which use uranium-235 (0.7% of all natural uranium), fast breeder reactors use uranium-238 (99.3% of all natural uranium). It has been estimated that there is up to five billion years' worth of uranium-238 for use in these power plants.

Breeder technology has been used in several reactors, but the high cost of reprocessing fuel safely, at 2006 technological levels, requires uranium prices of more than 200 USD/kg before becoming justified economically. Breeder reactors are still however being pursued as they have the potential to burn up all of the actinides in the present inventory of nuclear waste while also producing power and creating additional quantities of fuel for more reactors via the breeding process. In 2005, there were two breeder reactors producing power: the Phénix in France, which has since powered down in 2009 after 36 years of operation, and the BN-600 reactor, a reactor constructed in 1980 Beloyarsk, Russia which is still operational as of 2013. The electricity output of BN-600 is 600 MW — Russia plans to expand the nation's use of breeder reactors with the BN-800 reactor, was scheduled to become operational in 2014, but due to delays is not scheduled to produce power until 2017. The technical design of a yet larger breeder, the BN-1200 reactor was originally scheduled to be finalized in 2013, with construction slated for 2015 but has also been delayed. Japan's Monju breeder reactor restarted (having been shut down in 1995) in 2010 for 3 months, but shut down again after equipment fell into the reactor during reactor checkups, it is planned to become re-operational in late 2013. Both China and India are building breeder reactors. With the Indian 500 MWe Prototype Fast Breeder Reactor scheduled to become operational in 2014, with plans to build five more by 2020. The China Experimental Fast Reactor began producing power in 2011.

Another alternative to fast breeders is thermal breeder reactors that use uranium-233 bred from thorium as fission fuel in the thorium fuel cycle. Thorium is about 3.5 times more common than uranium in the Earth's crust, and has different geographic characteristics. This would extend the total practical fissionable resource base by 450%. India's three-stage nuclear power programme features the use of a thorium fuel cycle in the third stage, as it has abundant thorium reserves but little uranium.

Solid Waste

The most important waste stream from nuclear power plants is spent nuclear fuel. It is primarily composed of unconverted uranium as well as significant quantities of transuranic actinides (plutonium and curium, mostly). In addition, about 3% of it is fission products from nuclear reactions. The actinides (uranium, plutonium, and curium) are responsible for the bulk of the long-term radioactivity, whereas the fission products are responsible for the bulk of the short-term radioactivity.

High-level Radioactive Waste

A nuclear fuel rod assembly bundle being inspected before entering a reactor.

Following interim storage in a spent fuel pool, the bundles of used fuel assemblies of a typical nuclear power station are often stored on site in the likes of the eight dry cask storage vessels pictured above. At Yankee Rowe Nuclear Power Station, which generated 44 billion kilowatt hours of electricity over its lifetime, its complete spent fuel inventory is contained within sixteen casks.

High-level radioactive waste management concerns management and disposal of highly radioactive materials created during production of nuclear power. The technical issues in accomplishing this are daunting, due to the extremely long periods radioactive wastes remain deadly to living organisms. Of particular concern are two long-lived fission products, Technetium-99 (half-life 220,000 years) and Iodine-129 (half-life 15.7 million years), which dominate spent nuclear fuel radioactivity after a few thousand years. The most troublesome transuranic elements in spent fuel are Neptunium-237 (half-life two million years) and Plutonium-239 (half-life 24,000 years). Consequently, high-level radioactive waste requires sophisticated treatment and management to successfully isolate it from the biosphere. This usually necessitates treatment, followed by a long-term management strategy involving permanent storage, disposal or transformation of the waste into a non-toxic form.

Governments around the world are considering a range of waste management and disposal options, usually involving deep-geologic placement, although there has been limited progress toward implementing long-term waste management solutions. This is partly because the timeframes in question when dealing with radioactive waste range from 10,000 to millions of years, according to studies based on the effect of estimated radiation doses.

Some proposed nuclear reactor designs however such as the American Integral Fast Reactor and the Molten salt reactor can use the nuclear waste from light water reactors as a fuel, transmutating it to isotopes that would be safe after hundreds, instead of tens of thousands of years. This offers a potentially more attractive alternative to deep geological disposal.

Another possibility is the use of thorium in a reactor especially designed for thorium (rather than mixing in thorium with uranium and plutonium (i.e. in existing reactors). Used thorium fuel remains only a few hundreds of years radioactive, instead of tens of thousands of years.

Since the fraction of a radioisotope's atoms decaying per unit of time is inversely proportional to its half-life, the relative radioactivity of a quantity of buried human radioactive waste would diminish over time compared to natural radioisotopes (such as the decay chains of 120 trillion tons of thorium and 40 trillion tons of uranium which are at relatively trace concentrations of parts per million each over the crust's $3 * 10^{19}$ ton mass). For instance, over a timeframe of thousands of years, after the most active short half-life radioisotopes decayed, burying U.S. nuclear waste would increase the radioactivity in the top 2000 feet of rock and soil in the United States (10 million km²) by ≈ 1 part in 10 million over the cumulative amount of natural radioisotopes in such a volume, although the vicinity of the site would have a far higher concentration of artificial radioisotopes underground than such an average.

Low-level Radioactive Waste

The nuclear industry also produces a large volume of low-level radioactive waste in the form of contaminated items like clothing, hand tools, water purifier resins, and (upon decommissioning) the materials of which the reactor itself is built. In the US, the Nuclear Regulatory Commission has repeatedly attempted to allow low-level materials to be handled as normal waste: landfilled, recycled into consumer items, etcetera.

Comparing Radioactive Waste to Industrial Toxic Waste

In countries with nuclear power, radioactive wastes comprise less than 1% of total industrial toxic wastes, much of which remains hazardous for long periods. Overall, nuclear power produces far less waste material by volume than fossil-fuel based power plants. Coal-burning plants are particularly noted for producing large amounts of toxic and mildly radioactive ash due to concentrating naturally occurring metals and mildly radioactive material from the coal. A 2008 report from Oak Ridge National Laboratory concluded that coal power actually results in more radioactivity being released into the environment than nuclear power operation, and that the population effective dose equivalent, or dose to the public from radiation from coal plants is 100 times as much as from the ideal operation of nuclear plants. Indeed, coal ash is much less radioactive than spent nuclear fuel on a weight per weight basis, but coal ash is produced in much higher quantities per unit of energy generated, and this is released directly into the environment as fly ash, whereas nuclear plants use shielding to protect the environment from radioactive materials, for example, in dry cask storage vessels.

Waste Disposal

Disposal of nuclear waste is often said to be the Achilles' heel of the industry. Presently, waste is mainly stored at individual reactor sites and there are over 430 locations around the world where radioactive material continues to accumulate. Some experts suggest that centralized underground repositories which are well-managed, guarded, and monitored, would be a vast improvement. There is an "international consensus on the advisability of storing nuclear waste in deep geological repositories", with the lack of movement of nuclear waste in the 2 billion year old natural nuclear fission reactors in Oklo, Gabon being cited as "a source of essential information today."

There are no commercial scale purpose built underground repositories in operation. The Waste Isolation Pilot Plant (WIPP) in New Mexico has been taking nuclear waste since 1999 from production reactors, but as the name suggests is a research and development facility. A radiation leak at WIPP in 2014 brought renewed attention to the need for R&D on disposal or radioactive waste and spent fuel.

Reprocessing

Reprocessing can potentially recover up to 95% of the remaining uranium and plutonium in spent nuclear fuel, putting it into new mixed oxide fuel. This produces a reduction in long term radioactivity within the remaining waste, since this is largely short-lived fission products, and reduces its volume by over 90%. Reprocessing of civilian fuel from power reactors is currently done in Europe, Russia, Japan, and India. The full potential of reprocessing has not been achieved because it requires breeder reactors, which are not commercially available.

Nuclear reprocessing reduces the volume of high-level waste, but by itself does not reduce radioactivity or heat generation and therefore does not eliminate the need for a geological waste repository. Reprocessing has been politically controversial because of the potential to contribute to nuclear proliferation, the potential vulnerability to nuclear terrorism, the political challenges of repository siting (a problem that applies equally to direct disposal of spent fuel), and because of its high cost compared to the once-through fuel cycle. Several different methods for reprocessing been tried, but many have had safety and practicality problems which have led to their discontinuation.

In the United States, the Obama administration stepped back from President Bush's plans for commercial-scale reprocessing and reverted to a program focused on reprocessing-related scientific research. Reprocessing is not allowed in the U.S. In the U.S., spent nuclear fuel is currently all treated as waste. A major recommendation of the Blue Ribbon Commission on America's Nuclear Future was that "the United States should undertake an integrated nuclear waste management program that leads to the timely development of one or more permanent deep geological facilities for the safe disposal of spent fuel and high-level nuclear waste".

Depleted Uranium

Uranium enrichment produces many tons of depleted uranium (DU) which consists of U-238 with most of the easily fissile U-235 isotope removed. U-238 is a tough metal with several commercial uses—for example, aircraft production, radiation shielding, and armor—as it has a higher density than lead. Depleted uranium is also controversially used in munitions; DU penetrators (bullets or APFSDS tips) "self sharpen", due to uranium's tendency to fracture along shear bands.

Accidents, Attacks and Safety

Accidents

The 2011 Fukushima Daiichi nuclear disaster, the world's worst nuclear accident since 1986, displaced 50,000 households after radiation leaked into the air, soil and sea. Radiation checks led to bans of some shipments of vegetables and fish.

Some serious nuclear and radiation accidents have occurred. Benjamin K. Sovacool has reported that worldwide there have been 99 accidents at nuclear power plants. Fifty-seven accidents have occurred since the Chernobyl disaster, and 57% (56 out of 99) of all nuclear-related accidents have occurred in the USA.

Nuclear power plant accidents include the Chernobyl accident (1986) with approximately 60 deaths so far attributed to the accident and a predicted, eventual total death toll, of from 4000 to 25,000 latent cancers deaths. The Fukushima Daiichi nuclear disaster (2011), has not caused any radiation related deaths, with a predicted, eventual total death toll, of from 0 to 1000, and the Three Mile Island accident (1979), no causal deaths, cancer or otherwise, have been found in follow up studies of this accident. Nuclear-powered submarine mishaps include the K-19 reactor accident (1961), the K-27 reactor accident (1968), and the K-431 reactor accident (1985). International research is continuing into safety improvements such as passively safe plants, and the possible future use of nuclear fusion.

In terms of lives lost per unit of energy generated, nuclear power has caused fewer accidental deaths per unit of energy generated than all other major sources of energy generation. Energy produced by coal, petroleum, natural gas and hydropower has caused more deaths per unit of energy generated, from air pollution and energy accidents. This is found in the following comparisons, when the immediate nuclear related deaths from accidents are compared to the immediate deaths from these other energy sources, when the latent, or predicted, indirect cancer deaths from nuclear energy accidents are compared to the immediate deaths from the above energy sources, and when the combined immediate and indirect fatalities from nuclear power and all fossil fuels are compared, fatalities resulting from the mining of the necessary natural resources to power generation and to air pollution. With these data, the use of nuclear power has been calculated to have prevented in the region of 1.8 million deaths between 1971 and 2009, by reducing the proportion of energy that would otherwise have been generated by fossil fuels, and is projected to continue to do so.

Although according to Benjamin K. Sovacool, fission energy accidents ranked first in terms of their total economic cost, accounting for 41 percent of all property damage attributed to energy accidents. Analysis presented in the international Journal, *Human and Ecological Risk Assessment* found that coal, oil, Liquid petroleum gas and hydroelectric accidents(primarily due to the Banqiao dam burst) have resulted in greater economic impacts than nuclear power accidents.

Following the 2011 Japanese Fukushima nuclear disaster, authorities shut down the nation's 54 nuclear power plants, but it has been estimated that if Japan had never adopted nuclear power, accidents and pollution from coal or gas plants would have caused more lost years of life. As of 2013, the Fukushima site remains highly radioactive, with some 160,000 evacuees still living in temporary housing, and some land will be unfarmable for centuries. The difficult Fukushima disaster cleanup will take 40 or more years, and cost tens of billions of dollars.

Forced evacuation from a nuclear accident may lead to social isolation, anxiety, depression, psychosomatic medical problems, reckless behavior, even suicide. Such was the outcome of the 1986 Chernobyl nuclear disaster in Ukraine. A comprehensive 2005 study concluded that "the mental health impact of Chernobyl is the largest public health problem unleashed by the accident to date". Frank N. von Hippel, a U.S. scientist, commented on the 2011 Fukushima nuclear disaster, saying that "fear of ionizing radiation could have long-term psychological effects on a large portion of the population in the contaminated areas". A 2015 report in *Lancet* explained that serious impacts of nuclear accidents were often not directly attributable to radiation exposure, but rather social and psychological effects. Evacuation and long-term displacement of affected populations created problems for many people, especially the elderly and hospital patients.

Attacks and Sabotage

Terrorists could target nuclear power plants in an attempt to release radioactive contamination into the community. The United States 9/11 Commission has said that nuclear power plants were potential targets originally considered for the September 11, 2001 attacks. An attack on a reactor's spent fuel pool could also be serious, as these pools are less protected than the reactor core. The release of radioactivity could lead to thousands of near-term deaths and greater numbers of long-term fatalities.

If nuclear power use is to expand significantly, nuclear facilities will have to be made extremely safe from attacks that could release massive quantities of radioactivity into the community. New reactor designs have features of passive safety, such as the flooding of the reactor core without active intervention by reactor operators. But these safety measures have generally been developed and studied with respect to accidents, not to the deliberate reactor attack by a terrorist group. However, the US Nuclear Regulatory Commission does now also require new reactor license applications to consider security during the design stage. In the United States, the NRC carries out "Force on Force" (FOF) exercises at all Nuclear Power Plant (NPP) sites at least once every three years. In the U.S., plants are surrounded by a double row of tall fences which are electronically monitored. The plant grounds are patrolled by a sizeable force of armed guards.

Insider sabotage regularly occurs, because insiders can observe and work around security measures. Successful insider crimes depended on the perpetrators' observation and knowledge of security vulnerabilities. A fire caused 5–10 million dollars worth of damage to New York's Indian Point Energy Center in 1971. The arsonist turned out to be a plant maintenance worker. Sabotage by workers has been reported at many other reactors in the United States: at Zion Nuclear Power Station (1974), Quad Cities Nuclear Generating Station, Peach Bottom Nuclear Generating Station, Fort St. Vrain Generating Station, Trojan Nuclear Power Plant (1974), Browns Ferry Nuclear Power Plant (1980), and Beaver Valley Nuclear Generating Station (1981). Many reactors overseas have also reported sabotage by workers.

Nuclear Proliferation

Many technologies and materials associated with the creation of a nuclear power program have a dual-use capability, in that they can be used to make nuclear weapons if a country chooses to do so. When this happens a nuclear power program can become a route leading to a nuclear weapon or a public annex to a "secret" weapons program. The concern over Iran's nuclear activities is a case in point.

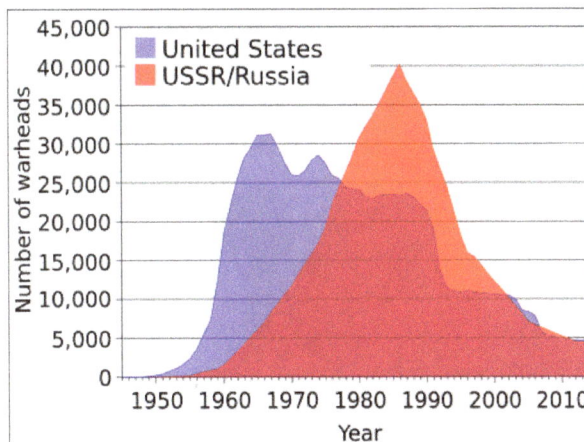

United States and USSR/Russian nuclear weapons stockpiles, 1945-2006. The Megatons to Mega-watts Program was the main driving force behind the sharp reduction in the quantity of nuclear weapons worldwide since the cold war ended. However without an increase in nuclear reactors and greater demand for fissile fuel, the cost of dismantling has dissuaded Russia from continuing their disarmament.

A fundamental goal for American and global security is to minimize the nuclear proliferation risks associated with the expansion of nuclear power. If this development is "poorly managed or efforts to contain risks are unsuccessful, the nuclear future will be dangerous". The Global Nuclear Energy Partnership is one such international effort to create a distribution network in which developing countries in need of energy, would receive nuclear fuel at a discounted rate, in exchange for that nation agreeing to forgo their own indigenous develop of a uranium enrichment program. The France-based Eurodif/*European Gaseous Diffusion Uranium Enrichment Consortium* was/is one such program that successfully implemented this concept, with Spain and other countries without enrichment facilities buying a share of the fuel produced at the French controlled enrichment facility, but without a transfer of technology. Iran was an early participant from 1974, and remains a shareholder of Eurodif via Sofidif.

According to Benjamin K. Sovacool, a "number of high-ranking officials, even within the United Nations, have argued that they can do little to stop states using nuclear reactors to produce nuclear weapons". A 2009 United Nations report said that:

the revival of interest in nuclear power could result in the worldwide dissemination of uranium enrichment and spent fuel reprocessing technologies, which present obvious risks of proliferation as these technologies can produce fissile materials that are directly usable in nuclear weapons.

On the other hand, one factor influencing the support of power reactors is due to the appeal that these reactors have at reducing nuclear weapons arsenals through the Megatons to Megawatts Program, a program which eliminated 425 metric tons of highly enriched uranium(HEU), the equivalent of 17,000 nuclear warheads, by diluting it with natural uranium making it equivalent to low enriched uranium(LEU), and thus suitable as nuclear fuel for commercial fission reactors. This is the single most successful non-proliferation program to date.

How much usable nuclear energy is placed in 1 (Minuteman) ICBM ?

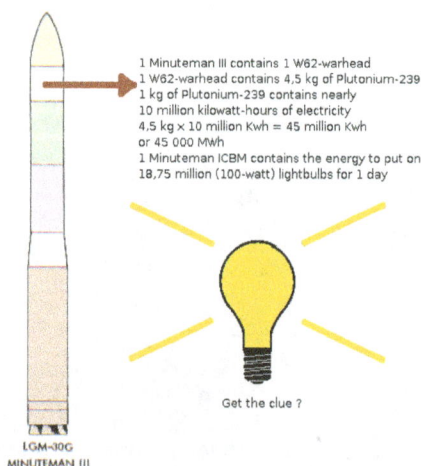

1 Minuteman III contains 1 W62-warhead
1 W62-warhead contains 4,5 kg of Plutonium-239
1 kg of Plutonium-239 contains nearly
10 million kilowatt-hours of electricity
4,5 kg x 10 million Kwh = 45 million Kwh
or 45 000 MWh
1 Minuteman ICBM contains the energy to put on
18,75 million (100-watt) lightbulbs for 1 day

Get the clue ?

LGM-30G
MINUTEMAN III

The Megatons to Megawatts Program, the brainchild of Thomas Neff of MIT, was hailed as a major success by anti-nuclear weapon advocates as it has largely been the driving force behind the sharp reduction in the quantity of nuclear weapons worldwide since the cold war ended. However without an increase in nuclear reactors and greater demand for fissile fuel, the cost of dismantling and down blending has dissuaded Russia from continuing their disarmament.

Currently, according to Harvard professor Matthew Bunn: "The Russians are not remotely interested in extending the program beyond 2013. We've managed to set it up in a way that costs them more and profits them less than them just making new low-enriched uranium for reactors from scratch. But there are other ways to set it up that would be very profitable for them and would also serve some of their strategic interests in boosting their nuclear exports."

Up to 2005, the Megatons to Megawatts Program had processed $8 billion of HEU/weapons grade uranium into LEU/reactor grade uranium, with that corresponding to the elimination of 10,000 nuclear weapons.

For approximately two decades, this material generated nearly 10 percent of all the electricity consumed in the United States (about half of all US nuclear electricity generated) with a total of around 7 trillion kilowatt-hours of electricity produced. Enough energy to energize the entire United States electric grid for about two years. In total it is estimated to have cost $17 billion, a "bargain for US ratepayers", with Russia profiting $12 billion from the deal. Much needed profit for the Russian nuclear oversight industry, which after the collapse of the Soviet economy, had difficulties paying for the maintenance and security of the Russian Federations highly enriched uranium and warheads.

In April 2012 there were thirty one countries that have civil nuclear power plants, of which nine have nuclear weapons, with the vast majority of these nuclear weapons states having first produced weapons, before commercial fission electricity stations. Moreover, the re-purposing of civilian nuclear industries for military purposes would be a breach of the Non-proliferation treaty, of which 190 countries adhere to.

Environmental Issues

Carbon emissions from nuclear power
Sovacool life cycle study survey, 2008

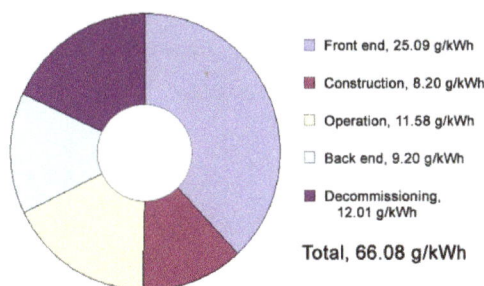

- Front end, 25.09 g/kWh
- Construction, 8.20 g/kWh
- Operation, 11.58 g/kWh
- Back end, 9.20 g/kWh
- Decommissioning, 12.01 g/kWh

Total, 66.08 g/kWh

Mean value of carbon dioxide emissions from qualified life cycle studies among 103 surveyed. Includes results of 1997 Vattenfall study.

A 2008 synthesis of 103 studies, published by Benjamin K. Sovacool, estimated that the value of CO_2 emissions for nuclear power over the lifecycle of a plant was 66.08 g/kW·h. Comparative results for various renewable power sources were 9–32 g/kW·h. A 2012 study by Yale University arrived at a different value, with the mean value, depending on which Reactor design was analyzed, ranging from 11 to 25 g/kW·h of total life cycle nuclear power CO_2 emissions.

Life cycle analysis (LCA) of carbon dioxide emissions show nuclear power as comparable to renewable energy sources. Emissions from burning fossil fuels are many times higher.

According to the United Nations (UNSCEAR), regular nuclear power plant operation including the nuclear fuel cycle causes radioisotope releases into the environment amounting to 0.0002 millisieverts (mSv) per year of public exposure as a global average. (Such is small compared to variation in natural background radiation, which averages 2.4 mSv/a globally but frequently varies between 1 mSv/a and 13 mSv/a depending on a person's location as determined by UNSCEAR). As of a 2008 report, the remaining legacy of the worst nuclear power plant accident (Chernobyl) is 0.002 mSv/a in global average exposure (a figure which was 0.04 mSv per person averaged over the entire populace of the Northern Hemisphere in the year of the accident in 1986, although far higher among the most affected local populations and recovery workers).

Climate Change

Climate change causing weather extremes such as heat waves, reduced precipitation levels and droughts can have a significant impact on nuclear energy infrastructure. Seawater is corrosive and so nuclear energy supply is likely to be negatively affected by the fresh water shortage. This generic problem may become increasingly significant over time. This can force nuclear reactors to be shut down, as happened in France during the 2003 and 2006 heat waves. Nuclear power supply was severely diminished by low river flow rates and droughts, which meant rivers had reached the maximum temperatures for cooling reactors. During the heat waves, 17 reactors had to limit output or shut down. 77% of French electricity is produced by nuclear power and in 2009 a similar situation created a 8GW shortage and forced the French government to import electricity. Other cases have been reported from Germany, where extreme temperatures have reduced nuclear power production 9 times due to high temperatures between 1979 and 2007. In particular:

- the Unterweser nuclear power plant reduced output by 90% between June and September 2003

- the Isar nuclear power plant cut production by 60% for 14 days due to excess river temperatures and low stream flow in the river Isar in 2006

Similar events have happened elsewhere in Europe during those same hot summers. If global warming continues, this disruption is likely to increase.

Comparison with Renewable Energy

As of 2013, the World Nuclear Association has said "There is unprecedented interest in renewable energy, particularly solar and wind energy, which provide electricity without giving rise to any carbon dioxide emission. Harnessing these for electricity depends on the cost and efficiency of the technology, which is constantly improving, thus reducing costs per peak kilowatt".

Renewable electricity production, from sources such as wind power and solar power, is sometimes criticized for being intermittent or variable. The International Energy Agency concluded that deployment of renewable technologies (RETs), when it increases the diversity of electricity sources, contributes to the flexibility of the system. Their report also concluded: "At high levels of grid penetration by RETs the consequences of unmatched demand and supply can pose challenges for grid management. This characteristic may affect how, and the degree to which, RETs can displace fossil fuels and nuclear capacities in power generation."

Renewable electricity supply in the 20-50+% range has already been implemented in several European systems, albeit in the context of an integrated European grid system. In 2012, the share of electricity generated by renewable sources in Germany was 21.9%, compared to 16.0% for nuclear power after Germany shut down 7-8 of its 18 nuclear reactors in 2011. In the United Kingdom, the amount of energy produced from renewable energy is expected to exceed that from nuclear power by 2018, and Scotland plans to obtain all electricity from renewable energy by 2020. The majority of installed renewable energy across the world is in the form of hydro power.

The IPCC has said that if governments were supportive, and the full complement of renewable energy technologies were deployed, renewable energy supply could account for almost 80% of the world's energy use within forty years. Rajendra Pachauri, chairman of the IPCC, said the necessary investment in renewables would cost only about 1% of global GDP annually. This approach could contain greenhouse gas levels to less than 450 parts per million, the safe level beyond which climate change becomes catastrophic and irreversible.

The cost of nuclear power has followed an increasing trend whereas the cost of electricity is declining for wind power. In about 2011, wind power became as inexpensive as natural gas, and anti-nuclear groups have suggested that in 2010 solar power became cheaper than nuclear power. Data from the EIA in 2011 estimated that in 2016, solar will have a levelized cost of electricity almost twice that of nuclear (21¢/kWh for solar, 11.39¢/kWh for nuclear), and wind somewhat less (9.7¢/kWh). However, the US EIA has also cautioned that levelized costs of intermittent sources such as wind and solar are not directly comparable to costs of "dispatchable" sources (those that can be adjusted to meet demand).

From a safety stand point, nuclear power, in terms of lives lost per unit of electricity delivered, is comparable to and in some cases, lower than many renewable energy sources. There is however no radioactive spent fuel that needs to be stored or reprocessed with conventional renewable energy sources. A nuclear plant needs to be disassembled and removed. Much of the disassembled nuclear plant needs to be stored as low level nuclear waste.

Nuclear Decommissioning

The price of energy inputs and the environmental costs of every nuclear power plant continue long after the facility has finished generating its last useful electricity. Once no longer economically viable, nuclear reactors and uranium enrichment facilities are generally decommissioned, returning the facility and its parts to a safe enough level to be entrusted for other uses, such as greenfield status. After a cooling-off period that may last decades, reactor core materials are dismantled and cut into small pieces to be packed in containers for interim storage or transmutation experiments. The process is expensive, time-consuming, dangerous for workers and potentially hazardous to the natural environment as it presents opportunities for human error, accidents or sabotage.

The total energy required for decommissioning can be as much as 50% more than the energy needed for the original construction. In most cases, the decommissioning process costs between US $300 million to US$5.6 billion. Decommissioning at nuclear sites which have experienced a serious accident are the most expensive and time-consuming. In the U.S. in 2011, there are

13 reactors that had permanently shut down and are in some phase of decommissioning. With Yankee Rowe Nuclear Power Station having completed the process in 2007, after ceasing commercial electricity production in 1992. The majority of the 15 years, was used to allow the station to naturally cool-down on its own, which makes the manual disassembly process both safer and cheaper.

Debate on Nuclear Power

The nuclear power debate concerns the controversy which has surrounded the deployment and use of nuclear fission reactors to generate electricity from nuclear fuel for civilian purposes. The debate about nuclear power peaked during the 1970s and 1980s, when it "reached an intensity unprecedented in the history of technology controversies", in some countries.

Proponents of nuclear energy contend that nuclear power is a sustainable energy source that reduces carbon emissions and increases energy security by decreasing dependence on imported energy sources. Proponents claim that nuclear power produces virtually no conventional air pollution, such as greenhouse gases and smog, in contrast to the chief viable alternative of fossil fuel. Nuclear power can produce base-load power unlike many renewables which are intermittent energy sources lacking large-scale and cheap ways of storing energy. M. King Hubbert saw oil as a resource that would run out, and proposed nuclear energy as a replacement energy source. Proponents claim that the risks of storing waste are small and can be further reduced by using the latest technology in newer reactors, and the operational safety record in the Western world is excellent when compared to the other major kinds of power plants.

Opponents believe that nuclear power poses many threats to people and the environment. These threats include the problems of processing, transport and storage of radioactive nuclear waste, the risk of nuclear weapons proliferation and terrorism, as well as health risks and environmental damage from uranium mining. They also contend that reactors themselves are enormously complex machines where many things can and do go wrong; and there have been serious nuclear accidents. Critics do not believe that the risks of using nuclear fission as a power source can be fully offset through the development of new technology. They also argue that when all the energy-intensive stages of the nuclear fuel chain are considered, from uranium mining to nuclear decommissioning, nuclear power is neither a low-carbon nor an economical electricity source.

Arguments of economics and safety are used by both sides of the debate.

Use in Space

Both fission and fusion appear promising for space propulsion applications, generating higher mission velocities with less reaction mass. This is due to the much higher energy density of nuclear reactions: some 7 orders of magnitude (10,000,000 times) more energetic than the chemical reactions which power the current generation of rockets.

Radioactive decay has been used on a relatively small scale (few kW), mostly to power space missions and experiments by using radioisotope thermoelectric generators such as those developed at Idaho National Laboratory.

Research

Advanced Concepts

Current fission reactors in operation around the world are second or third generation systems, with most of the first-generation systems having been retired some time ago. Research into advanced generation IV reactor types was officially started by the Generation IV International Forum (GIF) based on eight technology goals, including to improve nuclear safety, improve proliferation resistance, minimize waste, improve natural resource utilization, the ability to consume existing nuclear waste in the production of electricity, and decrease the cost to build and run such plants. Most of these reactors differ significantly from current operating light water reactors, and are generally not expected to be available for commercial construction before 2030.

The nuclear reactors to be built at Vogtle are new AP1000 third generation reactors, which are said to have safety improvements over older power reactors. However, John Ma, a senior structural engineer at the NRC, is concerned that some parts of the AP1000 steel skin are so brittle that the "impact energy" from a plane strike or storm driven projectile could shatter the wall. Edwin Lyman, a senior staff scientist at the Union of Concerned Scientists, is concerned about the strength of the steel containment vessel and the concrete shield building around the AP1000.

The Union of Concerned Scientists has referred to the EPR (nuclear reactor), currently under construction in China, Finland and France, as the only new reactor design under consideration in the United States that "...appears to have the potential to be significantly safer and more secure against attack than today's reactors."

One disadvantage of any new reactor technology is that safety risks may be greater initially as reactor operators have little experience with the new design. Nuclear engineer David Lochbaum has explained that almost all serious nuclear accidents have occurred with what was at the time the most recent technology. He argues that "the problem with new reactors and accidents is twofold: scenarios arise that are impossible to plan for in simulations; and humans make mistakes". As one director of a U.S. research laboratory put it, "fabrication, construction, operation, and maintenance of new reactors will face a steep learning curve: advanced technologies will have a heightened risk of accidents and mistakes. The technology may be proven, but people are not".

Hybrid Nuclear Fusion-fission

Hybrid nuclear power is a proposed means of generating power by use of a combination of nuclear fusion and fission processes. The concept dates to the 1950s, and was briefly advocated by Hans Bethe during the 1970s, but largely remained unexplored until a revival of interest in 2009, due to delays in the realization of pure fusion. When a sustained nuclear fusion power plant is built, it has the potential to be capable of extracting all the fission energy that remains in spent fission fuel, reducing the volume of nuclear waste by orders of magnitude, and more importantly, eliminating all actinides present in the spent fuel, substances which cause security concerns.

Nuclear Fusion

Nuclear fusion reactions have the potential to be safer and generate less radioactive waste than fission. These reactions appear potentially viable, though technically quite difficult and have yet to be created on a scale that could be used in a functional power plant. Fusion power has been under theoretical and experimental investigation since the 1950s.

Construction of the ITER facility began in 2007, but the project has run into many delays and budget overruns. The facility is now not expected to begin operations until the year 2027 – 11 years after initially anticipated. A follow on commercial nuclear fusion power station, DEMO, has been proposed. There are also suggestions for a power plant based upon a different fusion approach, that of an inertial fusion power plant.

Fusion powered electricity generation was initially believed to be readily achievable, as fission-electric power had been. However, the extreme requirements for continuous reactions and plasma containment led to projections being extended by several decades. In 2010, more than 60 years after the first attempts, commercial power production was still believed to be unlikely before 2050.

Marine Energy

Marine energy or marine power (also sometimes referred to as ocean energy, ocean power, or marine and hydrokinetic energy) refers to the energy carried by ocean waves, tides, salinity, and ocean temperature differences. The movement of water in the world's oceans creates a vast store of kinetic energy, or energy in motion. This energy can be harnessed to generate electricity to power homes, transport and industries.

The term marine energy encompasses both wave power i.e. power from surface waves, and tidal power i.e. obtained from the kinetic energy of large bodies of moving water. Offshore wind power is not a form of marine energy, as wind power is derived from the wind, even if the wind turbines are placed over water.

The oceans have a tremendous amount of energy and are close to many if not most concentrated populations. Ocean energy has the potential of providing a substantial amount of new renewable energy around the world.

Global Potential

There is the potential to develop 20,000–80,000 terawatt-hours (TWh) of electricity generated by changes in ocean temperatures, salt content, movements of tides, currents, waves and swells

Global potential	
Form	**Annual generation**
Tidal energy	>300 TWh
Marine current power	>800 TWh
Osmotic power Salinity gradient	2,000 TWh
Ocean thermal energy Thermal gradient	10,000 TWh
Wave energy	8,000–80,000 TWh
Source: IEA-OES, Annual Report 2007	

Indonesia as archipelagic country with three quarter of the area is ocean, has 49 GW recognized potential ocean energy and has 727 GW theoretical potential ocean energy.

Forms of Ocean Energy

Renewable

The oceans represent a vast and largely untapped source of energy in the form of surface waves, fluid flow, salinity gradients, and thermal.

Marine and Hydrokinetic (MHK) or marine energy development in U.S. and international waters includes projects using the following devices:

- Wave power converters in open coastal areas with significant waves;

- Tidal turbines placed in coastal and estuarine areas;

- In-stream turbines in fast-moving rivers;

- Ocean current turbines in areas of strong marine currents;

- Ocean thermal energy converters in deep tropical waters.

Marine Current Power

Strong ocean currents are generated from a combination of temperature, wind, salinity, bathymetry, and the rotation of the Earth. The Sun acts as the primary driving force, causing winds and temperature differences. Because there are only small fluctuations in current speed and stream location with no changes in direction, ocean currents may be suitable locations for deploying energy extraction devices such as turbines.

Ocean currents are instrumental in determining the climate in many regions around the world. While little is known about the effects of removing ocean current energy, the impacts of removing current energy on the farfield environment may be a significant environmental concern. The typical turbine issues with blade strike, entanglement of marine organisms, and acoustic effects still exists; however, these may be magnified due to the presence of more diverse populations of marine organisms using ocean currents for migration purposes. Locations can be further offshore and therefore require longer power cables that could affect the marine environment with electromagnetic output.

Osmotic Power

At the mouth of rivers where fresh water mixes with salt water, energy associated with the salinity gradient can be harnessed using pressure-retarded reverse osmosis process and associated conversion technologies. Another system is based on using freshwater upwelling through a turbine immersed in seawater, and one involving electrochemical reactions is also in development.

Significant research took place from 1975 to 1985 and gave various results regarding the economy of PRO and RED plants. It is important to note that small-scale investigations into salinity power production take place in other countries like Japan, Israel, and the United States. In Europe the research is concentrated in Norway and the Netherlands, in both places small pilots are tested. Salinity gradient energy is the energy available from the difference in salt concentration between freshwater with saltwater. This energy source is not easy to understand, as it is not directly occurring in nature in the form of heat, waterfalls, wind, waves, or radiation.

Ocean Thermal Energy

Water typically varies in temperature from the surface warmed by direct sunlight to greater depths where sunlight cannot penetrate. This differential is greatest in tropical waters, making this technology most applicable in water locations. A fluid is often vaporized to drive a turbine that may generate electricity or produce desalinized water. Systems may be either open-cycle, closed-cycle, or hybrid.

Tidal Power

The energy from moving masses of water — a popular form of hydroelectric power generation. Tidal power generation comprises three main forms, namely: tidal stream power, tidal barrage power, and dynamic tidal power.

Wave Power

Solar energy from the Sun creates temperature differentials that result in wind. The interaction between wind and the surface of water creates waves, which are larger when there is a greater distance for them to build up. Wave energy potential is greatest between 30° and 60° latitude in both hemispheres on the west coast because of the global direction of wind. When evaluating wave energy as a technology type, it is important to distinguish between the four most common approaches: point absorber buoys, surface attenuators, oscillating water columns, and overtopping devices.

The wave energy sector is reaching a significant milestone in the development of the industry, with positive steps towards commercial viability being taken. The more advanced device developers are now progressing beyond single unit demonstration devices and are proceeding to array development and multi-megawatt projects. The backing of major utility companies is now manifesting itself through partnerships within the development process, unlocking further investment and, in some cases, international co-operation.

At a simplified level, wave energy technology can be located near-shore and offshore. Wave energy converters can also be designed for operation in specific water depth conditions: deep water, intermediate water or shallow water. The fundamental device design will be dependent on the location of the device and the intended resource characteristics.

Non-renewable

Petroleum and natural gas beneath the ocean floor are also sometimes considered a form of ocean energy. An ocean engineer directs all phases of discovering, extracting, and delivering offshore petroleum (via oil tankers and pipelines,) a complex and demanding task. Also centrally important is the development of new methods to protect marine wildlife and coastal regions against the undesirable side effects of offshore oil extraction.

Marine Energy Development

The UK is leading the way in wave and tidal (marine) power generation. The world's first marine energy test facility was established in 2003 to kick start the development of the marine energy industry in the UK. Based in Orkney, Scotland, the European Marine Energy Centre (EMEC) has supported the deployment of more wave and tidal energy devices than at any other single site in the world. The Centre was established with around £36 million of funding from the Scottish Government, Highlands and Islands Enterprise, the Carbon Trust, UK Government, Scottish Enterprise, the European Union and Orkney Islands Council, and is the only accredited wave and tidal test centre for marine renewable energy in the world, suitable for testing a number of full-scale devices simultaneously in some of the harshest weather conditions while producing electricity to the national grid.

Clients currently testing at the centre include Aquamarine Power, Pelamis Wave Power, Scottish-Power Renewables and Wello on the wave site, and Alstom (formerly Tidal Generation Ltd), AN-DRITZ HYDRO Hammerfest, Open Hydro, Scotrenewables Tidal Power, and Voith on the tidal site.

Beyond device testing, EMEC also provides a wide range of consultancy and research services, and is working closely with Marine Scotland to streamline the consenting process for marine energy developers. EMEC is at the forefront in the development of international standards for marine energy, and is forging alliances with other countries, exporting its knowledge around the world to stimulate the development of a global marine renewables industry.

Environmental Effects

Common environmental concerns associated with marine energy developments include:

- the risk of marine mammals and fish being struck by tidal turbine blades

- the effects of EMF and underwater noise emitted from operating marine energy devices

- the physical presence of marine energy projects and their potential to alter the behavior of marine mammals, fish, and seabirds with attraction or avoidance

- the potential effect on nearfield and farfield marine environment and processes such as sediment transport and water quality

The Tethys database provides access to scientific literature and general information on the potential environmental effects of marine energy.

Tidal Power

Tidal power, also called tidal energy, is a form of hydropower that converts the energy obtained from tides into useful forms of power, mainly electricity.

Although not yet widely used, tidal power has potential for future electricity generation. Tides are more predictable than wind energy and solar power. Among sources of renewable energy, tidal power has traditionally suffered from relatively high cost and limited availability of sites with sufficiently high tidal ranges or flow velocities, thus constricting its total availability. However, many recent technological developments and improvements, both in design (e.g. dynamic tidal power, tidal lagoons) and turbine technology (e.g. new axial turbines, cross flow turbines), indicate that the total availability of tidal power may be much higher than previously assumed, and that economic and environmental costs may be brought down to competitive levels.

Historically, tide mills have been used both in Europe and on the Atlantic coast of North America. The incoming water was contained in large storage ponds, and as the tide went out, it turned waterwheels that used the mechanical power it produced to mill grain. The earliest occurrences date from the Middle Ages, or even from Roman times. It was only in the 19th century that the process of using falling water and spinning turbines to create electricity was introduced in the U.S. and Europe.

The world's first large-scale tidal power plant is the Rance Tidal Power Station in France, which became operational in 1966. It was the largest tidal power station in terms of power output, before Sihwa Lake Tidal Power Station surpassed it. Total harvestable energy from tidal areas close to the coast is estimated to be around 1 terawatt worldwide.

Generation of Tidal Energy

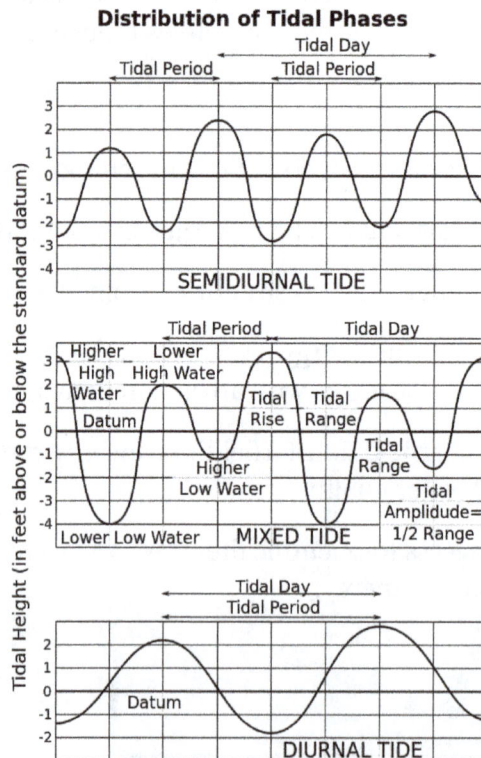

Variation of tides over a day

Tidal power is taken from the Earth's oceanic tides. Tidal forces are periodic variations in gravitational attraction exerted by celestial bodies. These forces create corresponding motions or currents in the world's oceans. Due to the strong attraction to the oceans, a bulge in the water level is created, causing a temporary increase in sea level. When the sea level is raised, water from the middle of the ocean is forced to move toward the shorelines, creating a tide. This occurrence takes place in an unfailing manner, due to the consistent pattern of the moon's orbit around the earth. The magnitude and character of this motion reflects the changing positions of the Moon and Sun relative to the Earth, the effects of Earth's rotation, and local geography of the sea floor and coastlines.

Tidal power is the only technology that draws on energy inherent in the orbital characteristics of the Earth–Moon system, and to a lesser extent in the Earth–Sun system. Other natural energies exploited by human technology originate directly or indirectly with the Sun, including fossil fuel, conventional hydroelectric, wind, biofuel, wave and solar energy. Nuclear energy makes use of Earth's mineral deposits of fissionable elements, while geothermal power taps the Earth's internal heat, which comes from a combination of residual heat from planetary accretion (about 20%) and heat produced through radioactive decay (80%).

A tidal generator converts the energy of tidal flows into electricity. Greater tidal variation and higher tidal current velocities can dramatically increase the potential of a site for tidal electricity generation.

Because the Earth's tides are ultimately due to gravitational interaction with the Moon and Sun and the Earth's rotation, tidal power is practically inexhaustible and classified as a renewable energy resource. Movement of tides causes a loss of mechanical energy in the Earth–Moon system: this is a result of pumping of water through natural restrictions around coastlines and consequent viscous dissipation at the seabed and in turbulence. This loss of energy has caused the rotation of the Earth to slow in the 4.5 billion years since its formation. During the last 620 million years the period of rotation of the earth (length of a day) has increased from 21.9 hours to 24 hours; in this period the Earth has lost 17% of its rotational energy. While tidal power will take additional energy from the system, the effect is negligible and would only be noticed over millions of years.

Generating Methods

The world's first commercial-scale and grid-connected tidal stream generator –
SeaGen – in Strangford Lough. The strong wake shows the power in the tidal current.

Tidal power can be classified into four generating methods:

Tidal Stream Generator

Tidal stream generators (or TSGs) make use of the kinetic energy of moving water to power turbines, in a similar way to wind turbines that use wind to power turbines. Some tidal generators can be built into the structures of existing bridges or are entirely submersed, thus avoiding concerns over impact on the natural landscape. Land constrictions such as straits or inlets can create high velocities at specific sites, which can be captured with the use of turbines. These turbines can be horizontal, vertical, open, or ducted and are typically placed near the bottom of the water column where tidal velocities are greatest.

Tidal Barrage

Tidal barrages make use of the potential energy in the difference in height (or hydraulic head) between high and low tides. When using tidal barrages to generate power, the potential energy from a tide is seized through strategic placement of specialized dams. When the sea level rises and the tide begins to come in, the temporary increase in tidal power is channeled into a large basin behind the dam, holding a large amount of potential energy. With the receding tide, this energy is then converted into mechanical energy as the water is released through large turbines that create electrical power through the use of generators. Barrages are essentially dams across the full width of a tidal estuary.

Dynamic Tidal Power

Top-down view of a DTP dam. Blue and dark red colors indicate low and high tides, respectively.

Dynamic tidal power (or DTP) is an untried but promising technology that would exploit an interaction between potential and kinetic energies in tidal flows. It proposes that very long dams (for example: 30–50 km length) be built from coasts straight out into the sea or ocean, without enclosing an area. Tidal phase differences are introduced across the dam, leading to a significant water-level differential in shallow coastal seas – featuring strong coast-parallel oscillating tidal currents such as found in the UK, China, and Korea.

Tidal Lagoon

A newer tidal energy design option is to construct circular retaining walls embedded with turbines that can capture the potential energy of tides. The created reservoirs are similar to those of tidal barrages, except that the location is artificial and does not contain a preexisting ecosystem. The lagoons can also be in double (or triple) format without pumping or with pumping that will flatten out the power output. The pumping power could be provided by excess to grid demand renewable energy from for example wind turbines or solar photovoltaic arrays. Excess renewable energy rather than being curtailed could be used and stored for a later period of time. Geographically dispersed tidal lagoons with a time delay between peak production would also flatten out peak production providing near base load production though at a higher cost than some other alternatives such as district heating renewable energy storage. The proposed Tidal Lagoon Swansea Bay in Wales, United Kingdom would be the first tidal power station of this type once built.

Us and Canadian Studies in the Twentieth Century

The first study of large scale tidal power plants was by the US Federal Power Commission in 1924 which if built would have been located in the northern border area of the US state of Maine and the south eastern border area of the Canadian province of New Brunswick, with various dams, power-houses, and ship locks enclosing the Bay of Fundy and Passamaquoddy Bay. Nothing came of the study and it is unknown whether Canada had been approached about the study by the US Federal Power Commission.

In 1956, utility Nova Scotia Light and Power of Halifax commissioned a pair of studies into the feasibility of commercial tidal power development on the Nova Scotia side of the Bay of Fundy. The two studies, by Stone & Webster of Boston and by Montreal Engineering Company of Montreal independently concluded that millions of horsepower could be harnessed from Fundy but that development costs would be commercially prohibitive at that time.

There was also a report on the international commission in April 1961 entitled "Investigation of the International Passamaquoddy Tidal Power Project" produced by both the US and Canadian Federal Governments. According to benefit to costs ratios, the project was beneficial to the US but not to Canada. A highway system along the top of the dams was envisioned as well.

A study was commissioned by the Canadian, Nova Scotian and New Brunswick governments (Re-assessment of Fundy Tidal Power) to determine the potential for tidal barrages at Chignecto Bay and Minas Basin – at the end of the Fundy Bay estuary. There were three sites determined to be financially feasible: Shepody Bay (1550 MW), Cumberline Basin (1085 MW), and Cobequid Bay (3800 MW). These were never built despite their apparent feasibility in 1977.

Tidal Power Development in the UK

The world's first marine energy test facility was established in 2003 to start the development of the wave and tidal energy industry in the UK. Based in Orkney, Scotland, the European Marine Energy Centre (EMEC) has supported the deployment of more wave and tidal energy devices than at any other single site in the world. EMEC provides a variety of test sites in real sea conditions. Its grid connected tidal test site is located at the Fall of Warness, off the island of Eday, in a narrow channel which concentrates the tide as it flows between the Atlantic Ocean and North Sea. This area has a very strong tidal current, which can travel up to 4 m/s (8 knots) in spring tides. Tidal energy developers currently testing at the site include: Alstom (formerly Tidal Generation Ltd); ANDRITZ HYDRO Hammerfest; OpenHydro; Scotrenewables Tidal Power; Voith. The resource could be 4 TJ per year.

Current and Future Tidal Power Schemes

- The first tidal power station was the Rance tidal power plant built over a period of 6 years from 1960 to 1966 at La Rance, France. It has 240 MW installed capacity.

- 254 MW Sihwa Lake Tidal Power Plant in South Korea is the largest tidal power installation in the world. Construction was completed in 2011.

- The first tidal power site in North America is the Annapolis Royal Generating Station, Annapolis Royal, Nova Scotia, which opened in 1984 on an inlet of the Bay of Fundy. It has 20 MW installed capacity.

- The Jiangxia Tidal Power Station, south of Hangzhou in China has been operational since 1985, with current installed capacity of 3.2 MW. More tidal power is planned near the mouth of the Yalu River.

- The first in-stream tidal current generator in North America (Race Rocks Tidal Power Demonstration Project) was installed at Race Rocks on southern Vancouver Island in September 2006. The next phase in the development of this tidal current generator will be in Nova Scotia (Bay of Fundy).

- A small project was built by the Soviet Union at Kislaya Guba on the Barents Sea. It has 0.4 MW installed capacity. In 2006 it was upgraded with a 1.2MW experimental advanced orthogonal turbine.

- Jindo Uldolmok Tidal Power Plant in South Korea is a tidal stream generation scheme planned to be expanded progressively to 90 MW of capacity by 2013. The first 1 MW was installed in May 2009.

- A 1.2 MW SeaGen system became operational in late 2008 on Strangford Lough in Northern Ireland.

- The contract for an 812 MW tidal barrage near Ganghwa Island (South Korea) north-west of Incheon has been signed by Daewoo. Completion is planned for 2015.

- A 1,320 MW barrage built around islands west of Incheon is proposed by the South Korean government, with projected construction starting in 2017.

- The Scottish Government has approved plans for a 10MW array of tidal stream generators near Islay, Scotland, costing 40 million pounds, and consisting of 10 turbines – enough to power over 5,000 homes. The first turbine is expected to be in operation by 2013.

- The Indian state of Gujarat is planning to host South Asia's first commercial-scale tidal power station. The company Atlantis Resources planned to install a 50MW tidal farm in the Gulf of Kutch on India's west coast, with construction starting early in 2012.

- Ocean Renewable Power Corporation was the first company to deliver tidal power to the US grid in September, 2012 when its pilot TidGen system was successfully deployed in Cobscook Bay, near Eastport.

- In New York City, 30 tidal turbines will be installed by Verdant Power in the East River by 2015 with a capacity of 1.05MW.

- Construction of a 320 MW tidal lagoon power plant outside the city of Swansea in the UK was granted planning permission in June 2015 and work is expected to start in 2016. Once completed, it will generate over 500GWh of electricity per year, enough to power roughly 155,000 homes.

- A turbine project is being installed in Ramsey Sound in 2014.

Tidal Power Issues

Environmental Concerns

Tidal power can have effects on marine life. The turbines can accidentally kill swimming sea life with the rotating blades, although projects such as the one in Strangford feature a safety mechanism that turns off the turbine when marine animals approach. Some fish may no longer utilize the area if threatened with a constant rotating or noise-making object. Marine life is a huge factor when placing tidal power energy generators in the water and precautions are made to ensure that as many marine animals as possible will not be affected by it. The Tethys database provides access to scientific literature and general information on the potential environmental effects of tidal energy.

Tidal Turbines

The main environmental concern with tidal energy is associated with blade strike and entanglement of marine organisms as high speed water increases the risk of organisms being pushed near or through these devices. As with all offshore renewable energies, there is also a concern about how the creation of EMF and acoustic outputs may affect marine organisms. It should be noted that because these devices are in the water, the acoustic output can be greater than those created with offshore wind energy. Depending on the frequency and amplitude of sound generated by the tidal energy devices, this acoustic output can have varying effects on marine mammals (particularly those who echolocate to communicate and navigate in the marine environment, such as dolphins and whales). Tidal energy removal can also cause environmental concerns such as degrading farfield water quality and disrupting sediment processes. Depending on the size of the project, these effects can range from small traces of sediment building up near the tidal device to severely affecting nearshore ecosystems and processes.

Tidal Barrage

Installing a barrage may change the shoreline within the bay or estuary, affecting a large ecosystem that depends on tidal flats. Inhibiting the flow of water in and out of the bay, there may also be less flushing of the bay or estuary, causing additional turbidity (suspended solids) and less saltwater, which may result in the death of fish that act as a vital food source to birds and mammals. Migrating fish may also be unable to access breeding streams, and may attempt to pass through the turbines. The same acoustic concerns apply to tidal barrages. Decreasing shipping accessibility can become a socio-economic issue, though locks can be added to allow slow passage. However, the barrage may improve the local economy by increasing land access as a bridge. Calmer waters may also allow better recreation in the bay or estuary. In August 2004, a humpback whale swam through the open sluice gate of the Annapolis Royal Generating Station at slack tide, ending up trapped for several days before eventually finding its way out to the Annapolis Basin.

Tidal Lagoon

Environmentally, the main concerns are blade strike on fish attempting to enter the lagoon, acoustic output from turbines, and changes in sedimentation processes. However, all these effects are localized and do not affect the entire estuary or bay.

Corrosion

Salt water causes corrosion in metal parts. It can be difficult to maintain tidal stream generators due to their size and depth in the water. The use of corrosion-resistant materials such as stainless steels, high-nickel alloys, copper-nickel alloys, nickel-copper alloys and titanium can greatly reduce, or eliminate, corrosion damage.

Mechanical fluids, such as lubricants, can leak out, which may be harmful to the marine life nearby. Proper maintenance can minimize the amount of harmful chemicals that may enter the environment.

Fouling

The biological events that happen when placing any structure in an area of high tidal currents and high biological productivity in the ocean will ensure that the structure becomes an ideal substrate for the growth of marine organisms. In the references of the Tidal Current Project at Race Rocks in British Columbia this is documented. Several structural materials and coatings were tested by the Lester Pearson College divers to assist Clean Current in reducing fouling on the turbine and other underwater infrastructure.

Structural Health Monitoring

The high load factors resulting from the fact that water is 800 times denser than air and the predictable and reliable nature of tides compared with the wind makes tidal energy particularly attractive for Electric power generation. Condition monitoring is the key for exploiting it cost-efficiently.

References

- Turcotte, D. L.; Schubert, G. (2002), Geodynamics (2 ed.), Cambridge, England, UK: Cambridge University Press, pp. 136–137, ISBN 978-0-521-66624-4 Cite uses deprecated parameter |coauthors= (help)

- Tiwari, G. N.; Ghosal, M. K. (2005), Renewable Energy Resources: Basic Principles and Applications, Alpha Science, ISBN 1-84265-125-0

- Starke, Linda (2009). State Of The World 2009: Into a Warming World: a WorldWatch Institute Report on Progress Toward a Sustainable Society. WW Norton & Company. ISBN 978-0-393-33418-0.

- Bain, Alastair S.; et al. (1997). Canada enters the nuclear age: a technical history of Atomic Energy of Canada. Magill-Queen's University Press. p. ix. ISBN 0-7735-1601-8.

- Pfau, Richard (1984) No Sacrifice Too Great: The Life of Lewis L. Strauss University Press of Virginia, Charlottesville, Virginia, p. 187, ISBN 978-0-8139-1038-3

- David Bodansky (2004). Nuclear Energy: Principles, Practices, and Prospects. Springer. p. 32. ISBN 978-0-387-20778-0. Retrieved 2008-01-31.

- Kragh, Helge (1999). Quantum Generations: A History of Physics in the Twentieth Century. Princeton NJ:

Princeton University Press. p. 286. ISBN 0-691-09552-3.

- McKeown, William (2003). Idaho Falls: The Untold Story of America's First Nuclear Accident. Toronto: ECW Press. ISBN 978-1-55022-562-4.

- Bernard L. Cohen (1990). The Nuclear Energy Option: An Alternative for the 90s. New York: Plenum Press. ISBN 978-0-306-43567-6.

- Rüdig, Wolfgang, ed. (1990). Anti-nuclear Movements: A World Survey of Opposition to Nuclear Energy. Detroit, MI: Longman Current Affairs. p. 1. ISBN 0-8103-9000-0.

- Falk, Jim (1982). Global Fission: The Battle Over Nuclear Power. Melbourne: Oxford University Press. pp. 95–96. ISBN 978-0-19-554315-5.

- Uranium 2007 – Resources, Production and Demand. Nuclear Energy Agency, Organisation for Economic Co-operation and Development. 2008-06-10. ISBN 978-92-64-04766-2.

- Ojovan, M. I.; Lee, W.E. (2005). An Introduction to Nuclear Waste Immobilisation. Amsterdam: Elsevier Science Publishers. p. 315. ISBN 0-08-044462-8.

- National Research Council (1995). Technical Bases for Yucca Mountain Standards. Washington, D.C.: National Academy Press. p. 91. ISBN 0-309-05289-0.

- Blue Ribbon Commission on America's Nuclear Future. "Disposal Subcommittee Report to the Full Commission" (PDF). Retrieved 1 January 2016.

- "Tidal - Capturing tidal fluctuations with turbines, tidal barrages, or tidal lagoons". Tidal / Tethys. Pacific Northwest National Laboratory (PNNL). Retrieved 2 February 2016.

- "World's First APR-1400 Connected to Grid". Washington DC, USA: NEI (Nuclear Energy Institute). 21 January 2016. Retrieved 7 March 2016.

Tools and Technologies used in Green Energy

With more and more machinery relying on green energy technologies, greater effort has been made to provide efficient and adaptable devices. Some of them that are discussed in this chapter are solar cell, wind turbine and biomass briquettes. This chapter discusses the methods of green energy technologies in a critical manner providing key analysis to the subject matter.

Wind Turbine

A wind turbine is a device that converts the wind's kinetic energy into electrical power. The term appears to have been adopted from hydroelectric technology (rotary propeller). The technical description of a wind turbine is aerofoil-powered generator.

Offshore wind farm, using 5 MW turbines REpower 5M in the North Sea off the coast of Belgium.

As a result of over a millennium of windmill development and modern engineering, today's wind turbines are manufactured in a wide range of vertical and horizontal axis types. The smallest turbines are used for applications such as battery charging for auxiliary power for boats or caravans or to power traffic warning signs. Slightly larger turbines can be used for making contributions to a domestic power supply while selling unused power back to the utility supplier via the electrical grid. Arrays of large turbines, known as wind farms, are becoming an increasingly important source of renewable energy and are used by many countries as part of a strategy to reduce their reliance on fossil fuels.

History

James Blyth's electricity-generating wind turbine, photographed in 1891

Windmills were used in Persia (present-day Iran) about 500-900 A.D. The windwheel of Hero of Alexandria marks one of the first known instances of wind powering a machine in history. However, the first known practical windmills were built in Sistan, an Eastern province of Iran, from the 7th century. These "Panemone" were vertical axle windmills, which had long vertical drive shafts with rectangular blades. Made of six to twelve sails covered in reed matting or cloth material, these windmills were used to grind grain or draw up water, and were used in the gristmilling and sugarcane industries.

Windmills first appeared in Europe during the Middle Ages. The first historical records of their use in England date to the 11th or 12th centuries and there are reports of German crusaders taking their windmill-making skills to Syria around 1190. By the 14th century, Dutch windmills were in use to drain areas of the Rhine delta. Advanced wind mills were described by Croatian inventor Fausto Veranzio. In his book Machinae Novae (1595) he described vertical axis wind turbines with curved or V-shaped blades.

The first electricity-generating wind turbine was a battery charging machine installed in July 1887 by Scottish academic James Blyth to light his holiday home in Marykirk, Scotland. Some months later American inventor Charles F. Brush was able to build the first automatically operated wind turbine after consulting local University professors and colleagues Jacob S. Gibbs and Brinsley Coleberd and successfully getting the blueprints peer-reviewed for electricity production in Cleveland, Ohio. Although Blyth's turbine was considered uneconomical in the United Kingdom electricity generation by wind turbines was more cost effective in countries with widely scattered populations.

In Denmark by 1900, there were about 2500 windmills for mechanical loads such as pumps and mills, producing an estimated combined peak power of about 30 MW. The largest machines were on 24-meter (79 ft) towers with four-bladed 23-meter (75 ft) diameter rotors. By 1908 there were 72 wind-driven electric generators operating in the United States from 5 kW to 25 kW. Around the time of World War I, American windmill makers were producing 100,000 farm windmills each year, mostly for water-pumping.

The first automatically operated wind turbine, built in Cleveland in 1887 by Charles F. Brush.
It was 60 feet (18 m) tall, weighed 4 tons (3.6 metric tonnes) and powered a 12 kW generator.

By the 1930s, wind generators for electricity were common on farms, mostly in the United States where distribution systems had not yet been installed. In this period, high-tensile steel was cheap, and the generators were placed atop prefabricated open steel lattice towers.

A forerunner of modern horizontal-axis wind generators was in service at Yalta, USSR in 1931. This was a 100 kW generator on a 30-meter (98 ft) tower, connected to the local 6.3 kV distribution system. It was reported to have an annual capacity factor of 32 percent, not much different from current wind machines.

In the autumn of 1941, the first megawatt-class wind turbine was synchronized to a utility grid in Vermont. The Smith-Putnam wind turbine only ran for 1,100 hours before suffering a critical failure. The unit was not repaired, because of shortage of materials during the war.

The first utility grid-connected wind turbine to operate in the UK was built by John Brown & Company in 1951 in the Orkney Islands.

Despite these diverse developments, developments in fossil fuel systems almost entirely eliminated any wind turbine systems larger than supermicro size. In the early 1970s, however, anti-nuclear protests in Denmark spurred artisan mechanics to develop microturbines of 22 kW. Organizing owners into associations and co-operatives lead to the lobbying of the government and utilities and provided incentives for larger turbines throughout the 1980s and later. Local activists in Germany, nascent turbine manufacturers in Spain, and large investors in the United States in the early 1990s then lobbied for policies that stimulated the industry in those countries. Later companies formed in India and China. As of 2012, Danish company Vestas is the world's biggest wind-turbine manufacturer.

Resources

Nordex N117/2400 in Germany, a modern low-wind turbine.

Wind turbines at the Jepirachí Eolian Park in La Guajira, Colombia.

A quantitative measure of the wind energy available at any location is called the Wind Power Density (WPD). It is a calculation of the mean annual power available per square meter of swept area of a turbine, and is tabulated for different heights above ground. Calculation of wind power density includes the effect of wind velocity and air density. Color-coded maps are prepared for a particular area described, for example, as "Mean Annual Power Density at 50 Metres". In the United States, the results of the above calculation are included in an index developed by the National Renewable Energy Laboratory and referred to as "NREL CLASS". The larger the WPD, the higher it is rated by class. Classes range from Class 1 (200 watts per square meter or less at 50 m altitude) to Class 7 (800 to 2000 watts per square m). Commercial wind farms generally are sited in Class 3 or higher areas, although isolated points in an otherwise Class 1 area may be practical to exploit.

Wind turbines are classified by the wind speed they are designed for, from class I to class IV, with A or B referring to the turbulence.

Class	Avg Wind Speed (m/s)	Turbulence
IA	10	18%
IB	10	16%
IIA	8.5	18%
IIB	8.5	16%
IIIA	7.5	18%
IIIB	7.5	16%
IVA	6	18%
IVB	6	16%

Efficiency

Not all the energy of blowing wind can be used, but some small wind turbines are designed to work at low wind speeds.

Conservation of mass requires that the amount of air entering and exiting a turbine must be equal. Accordingly, Betz's law gives the maximal achievable extraction of wind power by a wind turbine as 16/27 (59.3%) of the total kinetic energy of the air flowing through the turbine.

The maximum theoretical power output of a wind machine is thus 0.59 times the kinetic energy of the air passing through the effective disk area of the machine. If the effective area of the disk is A, and the wind velocity v, the maximum theoretical power output P is:

$$P = 0.59 \frac{1}{2} \rho v^3 A$$

where ρ is air density

As wind is free (no fuel cost), wind-to-rotor efficiency (including rotor blade friction and drag) is one of many aspects impacting the final price of wind power. Further inefficiencies, such as gearbox losses, generator and converter losses, reduce the power delivered by a wind turbine. To protect components from undue wear, extracted power is held constant above the rated operating speed as theoretical power increases at the cube of wind speed, further reducing theoretical efficiency. In 2001, commercial utility-connected turbines deliver 75% to 80% of the Betz limit of power extractable from the wind, at rated operating speed.

Efficiency can decrease slightly over time due to wear. Analysis of 3128 wind turbines older than 10 years in Denmark showed that half of the turbines had no decrease, while the other half saw a production decrease of 1.2% per year.

Types

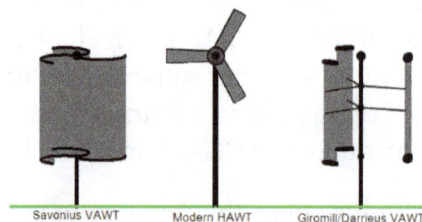

Savonius VAWT   Modern HAWT   Giromill/Darrieus VAWT

The three primary types: VAWT Savonius, HAWT towered; VAWT Darrieus as they appear in operation

Wind turbines can rotate about either a horizontal or a vertical axis, the former being both older and more common. They can also include blades (transparent or not) or be bladeless.

Horizontal Axis

Components of a horizontal axis wind turbine (gearbox, rotor shaft and brake assembly) being lifted into position

A turbine blade convoy passing through Edenfield, UK

Horizontal-axis wind turbines (HAWT) have the main rotor shaft and electrical generator at the top of a tower, and must be pointed into the wind. Small turbines are pointed by a simple wind vane, while large turbines generally use a wind sensor coupled with a servo motor. Most have a gearbox, which turns the slow rotation of the blades into a quicker rotation that is more suitable to drive an electrical generator.

Since a tower produces turbulence behind it, the turbine is usually positioned upwind of its supporting tower. Turbine blades are made stiff to prevent the blades from being pushed into the tower by high winds. Additionally, the blades are placed a considerable distance in front of the tower and are sometimes tilted forward into the wind a small amount.

Downwind machines have been built, despite the problem of turbulence (mast wake), because they don't need an additional mechanism for keeping them in line with the wind, and because in high winds the blades can be allowed to bend which reduces their swept area and thus their wind resistance. Since cyclical (that is repetitive) turbulence may lead to fatigue failures, most HAWTs are of upwind design.

Turbines used in wind farms for commercial production of electric power are usually three-bladed and pointed into the wind by computer-controlled motors. These have high tip speeds of over 320 km/h (200 mph), high efficiency, and low torque ripple, which contribute to good reliability. The blades are usually colored white for daytime visibility by aircraft and range in length from 20 to 40 meters (66 to 131 ft) or more. The tubular steel towers range from 60 to 90 meters (200 to 300 ft) tall.

The blades rotate at 10 to 22 revolutions per minute. At 22 rotations per minute the tip speed exceeds 90 meters per second (300 ft/s). A gear box is commonly used for stepping up the speed of the generator, although designs may also use direct drive of an annular generator. Some models operate at constant speed, but more energy can be collected by variable-speed turbines which use a solid-state power converter to interface to the transmission system. All turbines are equipped with protective features to avoid damage at high wind speeds, by feathering the blades into the wind which ceases their rotation, supplemented by brakes.

Year by year the size and height of turbines increase. Offshore wind turbines are built up to 8MW today and have a blade length up to 80m. Onshore wind turbines are installed in low wind speed areas and getting higher and higher towers. Usual towers of multi megawatt turbines have a height of 70 m to 120 m and in extremes up to 160 m, with blade tip speeds reaching 80 m/s to 90 m/s. Higher tip speeds means more noise and blade erosion.

Vertical Axis Design

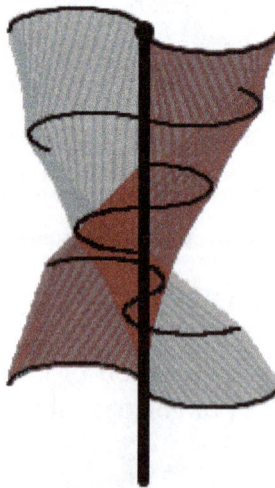

A vertical axis Twisted Savonius type turbine.

Vertical-axis wind turbines (or VAWTs) have the main rotor shaft arranged vertically. One advantage of this arrangement is that the turbine does not need to be pointed into the wind to be effective, which is an advantage on a site where the wind direction is highly variable. It is also an advantage when the turbine is integrated into a building because it is inherently less steerable. Also, the generator and gearbox can be placed near the ground, using a direct drive from the rotor assembly to the ground-based gearbox, improving accessibility for maintenance.

The key disadvantages include the relatively low rotational speed with the consequential higher torque and hence higher cost of the drive train, the inherently lower power coefficient, the 360-degree rotation of the aerofoil within the wind flow during each cycle and hence the highly dynamic loading on the blade, the pulsating torque generated by some rotor designs on the drive train, and the difficulty of modelling the wind flow accurately and hence the challenges of analysing and designing the rotor prior to fabricating a prototype.

When a turbine is mounted on a rooftop the building generally redirects wind over the roof and this can double the wind speed at the turbine. If the height of a rooftop mounted turbine tower is approximately 50% of the building height it is near the optimum for maximum wind energy and minimum wind turbulence. Wind speeds within the built environment are generally much lower than at exposed rural sites, noise may be a concern and an existing structure may not adequately resist the additional stress.

Subtypes of the vertical axis design include:

Offshore Horizontal Axis Wind Turbines (HAWTs) at Scroby Sands Wind Farm, UK

Onshore Horizontal Axis Wind Turbines in Zhangjiakou, China

Darrieus Wind Turbine

"Eggbeater" turbines, or Darrieus turbines, were named after the French inventor, Georges Darrieus. They have good efficiency, but produce large torque ripple and cyclical stress on the tower, which contributes to poor reliability. They also generally require some external power source, or an additional Savonius rotor to start turning, because the starting torque is very low. The torque ripple is reduced by using three or more blades which results in greater solidity of the rotor. Solidity is measured by blade area divided by the rotor area. Newer Darrieus type turbines are not held up by guy-wires but have an external superstructure connected to the top bearing.

Giromill

A subtype of Darrieus turbine with straight, as opposed to curved, blades. The cycloturbine variety has variable pitch to reduce the torque pulsation and is self-starting. The advantages of variable pitch are: high starting torque; a wide, relatively flat torque curve; a higher coefficient of performance; more efficient operation in turbulent winds; and a lower blade speed ratio which lowers blade bending stresses. Straight, V, or curved blades may be used.

Savonius Wind Turbine

These are drag-type devices with two (or more) scoops that are used in anemometers, *Flettner* vents (commonly seen on bus and van roofs), and in some high-reliability low-efficiency power turbines. They are always self-starting if there are at least three scoops.

Twisted Savonius

Twisted Savonius is a modified savonius, with long helical scoops to provide smooth torque. This is often used as a rooftop windturbine and has even been adapted for ships.

Another type of vertical axis is the Parallel turbine, which is similar to the crossflow fan or centrifugal fan. It uses the ground effect. Vertical axis turbines of this type have been tried for many years: a unit producing 10 kW was built by Israeli wind pioneer Bruce Brill in the 1980s.

Vortexis

The most recent advancement in Vertical Axis Wind Turbines has been the Vortexis VAWT, utilizing a pre-swirled augmented vertical axis wind turbine (PA-VAWT) designed for the purpose of developing a high efficiency VAWT concept that keeps the advantages of VAWT's compact size, lack of bias as to incoming wind direction, easy deployment and low radar cross section for use in mobile applications for the military, referred to in Special Operations as "Black Swan."

Design and Construction

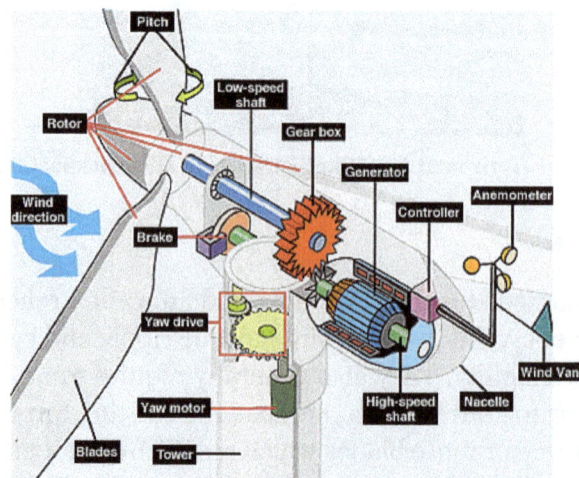

Components of a horizontal-axis wind turbine

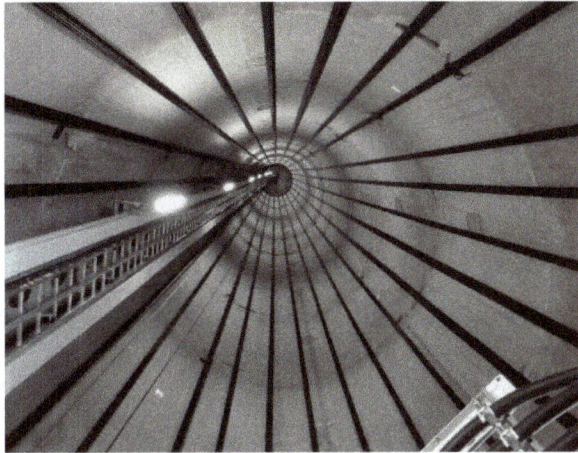

Inside view of a wind turbine tower, showing the tendon cables.

Wind turbines are designed to exploit the wind energy that exists at a location. Aerodynamic modeling is used to determine the optimum tower height, control systems, number of blades and blade shape.

Wind turbines convert wind energy to electricity for distribution. Conventional horizontal axis turbines can be divided into three components:

- The rotor component, which is approximately 20% of the wind turbine cost, includes the blades for converting wind energy to low speed rotational energy.

- The generator component, which is approximately 34% of the wind turbine cost, includes the electrical generator, the control electronics, and most likely a gearbox (e.g. planetary gearbox), adjustable-speed drive or continuously variable transmission component for converting the low speed incoming rotation to high speed rotation suitable for generating electricity.

- The structural support component, which is approximately 15% of the wind turbine cost, includes the tower and rotor yaw mechanism.

A 1.5 MW wind turbine of a type frequently seen in the United States has a tower 80 meters (260 ft) high. The rotor assembly (blades and hub) weighs 22,000 kilograms (48,000 lb). The nacelle, which contains the generator component, weighs 52,000 kilograms (115,000 lb). The concrete base for the tower is constructed using 26,000 kilograms (58,000 lb) of reinforcing steel and contains 190 cubic meters (250 cu yd) of concrete. The base is 15 meters (50 ft) in diameter and 2.4 meters (8 ft) thick near the center.

Among all renewable energy systems wind turbines have the highest effective intensity of power-harvesting surface because turbine blades not only harvest wind power, but also concentrate it.

Unconventional Designs

An E-66 wind turbine in the Windpark Holtriem, Germany, has an observation deck for visitors. Another turbine of the same type with an observation deck is located in Swaffham, England. Airborne wind turbine designs have been proposed and developed for many years but have yet to produce significant amounts of energy. In principle, wind turbines may also be used in conjunction with a large vertical solar updraft tower to extract the energy due to air heated by the sun.

The corkscrew shaped wind turbine at Progressive Field in Cleveland, Ohio

Wind turbines which utilise the Magnus effect have been developed.

A ram air turbine (RAT) is a special kind of small turbine that is fitted to some aircraft. When deployed, the RAT is spun by the airstream going past the aircraft and can provide power for the most essential systems if there is a loss of all on-board electrical power, as in the case of the "Gimli Glider".

The two-bladed turbine SCD 6MW offshore turbine designed by aerodyn Energiesysteme, built by MingYang Wind Power has a helideck for helicopters on top of its nacelle. The prototype was erected in 2014 in Rudong China.

Turbine Monitoring and Diagnostics

Due to data transmission problems, structural health monitoring of wind turbines is usually performed using several accelerometers and strain gages attached to the nacelle to monitor the gearbox and equipments. Currently, digital image correlation and stereophotogrammetry are used to measure dynamics of wind turbine blades. These methods usually measure displacement and strain to identify location of defects. Dynamic characteristics of non-rotating wind turbines have been measured using digital image correlation and photogrammetry. Three dimensional point tracking has also been used to measure rotating dynamics of wind turbines.

Materials and Durability

Currently serving wind turbine blades are mainly made of composite materials. These blades are usually made of a polyester resin, a vinyl resin, and epoxy thermosetting matrix resin and E-glass fibers, S- glass fibers and carbon fiber reinforced materials. Construction may use manual layup techniques or composite resin injection molding. As the price of glass fibers is only about one tenth the price of carbon fiber, glass fiber is still dominant. One of the predominant ways wind turbines have gain performance is by increasing rotor diameters, and thus blade length. Longer blades place more demands on the strength and stiffness of the materials. Stiffness is especially important to avoid having blades flex to the degree that they hit the tower of the wind turbine. Carbon fiber is between 4 and 6 times stiffer than glass fiber, so carbon fiber is becoming more common in wind turbine blades.

Wind Turbines on Public Display

The Nordex N50 wind turbine and visitor centre of Lamma Winds in Hong Kong, China

A few localities have exploited the attention-getting nature of wind turbines by placing them on public display, either with visitor centers around their bases, or with viewing areas farther away. The wind turbines are generally of conventional horizontal-axis, three-bladed design, and generate power to feed electrical grids, but they also serve the unconventional roles of technology demonstration, public relations, and education.

Small wind Turbines

A small Quietrevolution QR5 Gorlov type vertical axis wind turbine in Bristol, England.
Measuring 3 m in diameter and 5 m high, it has a nameplate rating of 6.5 kW to the grid.

Small wind turbines may be used for a variety of applications including on- or off-grid residences, telecom towers, offshore platforms, rural schools and clinics, remote monitoring and other purposes that require energy where there is no electric grid, or where the grid is unstable. Small wind turbines may be as small as a fifty-watt generator for boat or caravan use. Hybrid solar and wind powered units are increasingly being used for traffic signage, particularly in rural locations, as they avoid the need to lay long cables from the nearest mains connection point. The U.S. Department of Energy's National Renewable Energy Laboratory (NREL) defines small wind turbines as those smaller than or equal to 100 kilowatts. Small units often have direct drive generators, direct current output, aeroelastic blades, lifetime bearings and use a vane to point into the wind.

Larger, more costly turbines generally have geared power trains, alternating current output, flaps and are actively pointed into the wind. Direct drive generators and aeroelastic blades for large wind turbines are being researched.

Wind Turbine Spacing

On most horizontal windturbine farms, a spacing of about 6-10 times the rotor diameter is often upheld. However, for large wind farms distances of about 15 rotor diameters should be more economically optimal, taking into account typical wind turbine and land costs. This conclusion has been reached by research conducted by Charles Meneveau of the Johns Hopkins University, and Johan Meyers of Leuven University in Belgium, based on computer simulations that take into account the detailed interactions among wind turbines (wakes) as well as with the entire turbulent atmospheric boundary layer. Moreover, recent research by John Dabiri of Caltech suggests that vertical wind turbines may be placed much more closely together so long as an alternating pattern of rotation is created allowing blades of neighbouring turbines to move in the same direction as they approach one another.

Operability

Maintenance

Wind turbines need regular maintenance to stay reliable and available, reaching 98%.

Modern turbines usually have a small onboard crane for hoisting maintenance tools and minor components. However, large heavy components like generator, gearbox, blades and so on are rarely replaced and a heavy lift external crane is needed in those cases. If the turbine has a difficult access road, a containerized crane can be lifted up by the internal crane to provide heavier lifting.

Repowering

Installation of new wind turbines can be controversial. An alternative is repowering, where existing wind turbines are replaced with bigger, more powerful ones, sometimes in smaller numbers while keeping or increasing capacity.

Demolition

Older turbines were in some early cases not required to be removed when reaching the end of their life. Some still stand, waiting to be recycled or repowered.

A demolition industry develops to recycle offshore turbines at a cost of DKK 2–4 million per MW, to be guaranteed by the owner.

Records

Fuhrländer Wind Turbine Laasow, in Brandenburg, Germany, among the world's tallest wind turbines

Éole, the largest vertical axis wind turbine, in Cap-Chat, Quebec, Canada

Largest Capacity Conventional Drive

The Vestas V164 has a rated capacity of 8 MW, has an overall height of 220 m (722 ft), a diameter of 164 m (538 ft), is for offshore use, and is the world's largest-capacity wind turbine since its introduction in 2014. The conventional drive train consist of a main gearbox and a medium speed PM generator. Prototype installed in 2014 at the National Test Center Denmark nearby Østerild. Series production starts end of 2015.

Largest Capacity Direct Drive

The Enercon E-126 with 7.58 MW and 127 m rotor diameter is the largest direct drive turbine. It's only for onshore use. The turbine has parted rotor blades with 2 sections for transport. In July 2016, Siemens upgraded its 7 to 8 MW.

Largest Vertical-axis

Le Nordais wind farm in Cap-Chat, Quebec has a vertical axis wind turbine (VAWT) named Éole, which is the world's largest at 110 m. It has a nameplate capacity of 3.8 MW.

Largest 1-bladed Turbine

Riva Calzoni M33 was a single-bladed wind turbine with 350 kW, designed and built In Bologna in 1993.

Largest 2-bladed Turbine

Today's biggest 2-bladed turbine is build by Mingyang Wind Power in 2013. It is a SCD6.5MW offshore downwind turbine, designed by aerodyn Energiesysteme.

Largest Swept Area

The turbine with the largest swept area is the Samsung S7.0-171, with a diameter of 171 m, giving a total sweep of 22966 m².

Tallest

A Nordex 3.3 MW was installed in July 2016. It has a total height of 230m, and a hub height of 164m on 100m concrete tower bottom with steel tubes on top (hybrid tower).

Vestas V164 was the tallest wind turbine, standing in Østerild, Denmark, 220 meters tall, constructed in 2014. It has a steel tube tower.

Highest Tower

Fuhrländer installed a 2.5MW turbine on a 160m lattice tower in 2003.

Most Rotors

Lagerwey has build Four-in-One, a multi rotor wind turbine with one tower and four rotors near Maasvlakte. In April 2016, Vestas installed a 900 kW quadrotor test wind turbine at Risø, made from 4 recycled 225 kW V29 turbines.

Most Productive

Four turbines at Rønland wind farm in Denmark share the record for the most productive wind turbines, with each having generated 63.2 GWh by June 2010.

Highest-situated

Since 2013 the world's highest-situated wind turbine was made and installed by WindAid and is located at the base of the Pastoruri Glacier in Peru at 4,877 meters (16,001 ft) above sea level. The site uses the WindAid 2.5 kW wind generator to supply power to a small rural community of micro entrepreneurs who cater to the tourists who come to the Pastoruri glacier.

Largest Floating Wind Turbine

The world's largest—and also the first operational deep-water *large-capacity*—floating wind turbine is the 2.3 MW Hywind currently operating 10 kilometers (6.2 mi) offshore in 220-meter-deep water, southwest of Karmøy, Norway. The turbine began operating in September 2009 and utilizes a Siemens 2.3 MW turbine.

Solar Cell

A conventional crystalline silicon solar cell (as of 2005). Electrical contacts made from busbars (the larger silver-colored strips) and fingers (the smaller ones) are printed on the silicon wafer.

A solar cell, or photovoltaic cell (in very early days also termed "solar battery" – a denotation which nowadays has a totally different meaning), is an electrical device that converts the energy of light directly into electricity by the photovoltaic effect, which is a physical and chemical phenomenon. It is a form of photoelectric cell, defined as a device whose electrical characteristics, such as current, voltage, or resistance, vary when exposed to light. Solar cells are the building blocks of photovoltaic modules, otherwise known as solar panels.

Solar cells are described as being photovoltaic irrespective of whether the source is sunlight or an artificial light. They are used as a photodetector (for example infrared detectors), detecting light or other electromagnetic radiation near the visible range, or measuring light intensity.

The operation of a photovoltaic (PV) cell requires 3 basic attributes:

- The absorption of light, generating either electron-hole pairs or excitons.

- The separation of charge carriers of opposite types.

- The separate extraction of those carriers to an external circuit.

In contrast, a solar thermal collector supplies heat by absorbing sunlight, for the purpose of either direct heating or indirect electrical power generation from heat. A "photoelectrolytic cell" (photo-electrochemical cell), on the other hand, refers either to a type of photovoltaic cell (like that developed by Edmond Becquerel and modern dye-sensitized solar cells), or to a device that splits water directly into hydrogen and oxygen using only solar illumination.

Applications

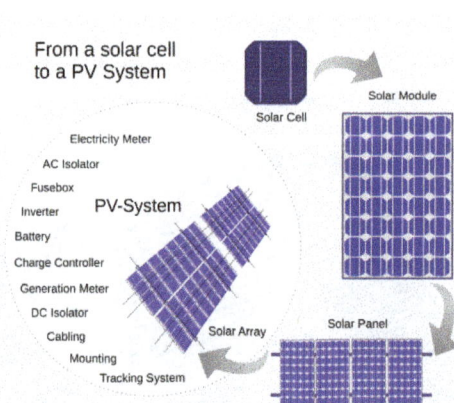

From a solar cell to a PV system. Diagram of the possible components of a photovoltaic system

Assemblies of solar cells are used to make solar modules which generate electrical power from sunlight, as distinguished from a "solar thermal module" or "solar hot water panel". A solar array generates solar power using solar energy.

Cells, Modules, Panels and Systems

Multiple solar cells in an integrated group, all oriented in one plane, constitute a solar photovoltaic panel or solar photovoltaic module. Photovoltaic modules often have a sheet of glass on the sun-facing side, allowing light to pass while protecting the semiconductor wafers. Solar cells are usually connected in series and parallel circuits or series in modules, creating an additive voltage. Connecting cells in parallel yields a higher current; however, problems such as shadow effects can shut down the weaker (less illuminated) parallel string (a number of series connected cells) causing substantial power loss and possible damage because of the reverse bias applied to the shadowed cells by their illuminated partners. Strings of series cells are usually handled independently and not connected in parallel, though as of 2014, individual power boxes are often supplied for each module, and are connected in parallel. Although modules can be interconnected to create an array with the desired peak DC voltage and loading current capacity, using independent MPPTs (maximum power point trackers) is preferable. Otherwise, shunt diodes can reduce shadowing power loss in arrays with series/parallel connected cells.

Typical PV system prices in 2013 in selected countries (USD)								
USD/W	Australia	China	France	Germany	Italy	Japan	United Kingdom	United States
Residential	1.8	1.5	4.1	2.4	2.8	4.2	2.8	4.9
Commercial	1.7	1.4	2.7	1.8	1.9	3.6	2.4	4.5
Utility-scale	2.0	1.4	2.2	1.4	1.5	2.9	1.9	3.3

Source: *IEA – Technology Roadmap: Solar Photovoltaic Energy report*, 2014 edition
Note: *DOE – Photovoltaic System Pricing Trends* reports lower prices for the U.S.

History

The photovoltaic effect was experimentally demonstrated first by French physicist Edmond Becquerel. In 1839, at age 19, he built the world's first photovoltaic cell in his father's laboratory. Willoughby Smith first described the "Effect of Light on Selenium during the passage of an Electric Current" in a 20 February 1873 issue of Nature. In 1883 Charles Fritts built the first solid state photovoltaic cell by coating the semiconductor selenium with a thin layer of gold to form the junctions; the device was only around 1% efficient.

In 1888 Russian physicist Aleksandr Stoletov built the first cell based on the outer photoelectric effect discovered by Heinrich Hertz in 1887.

In 1905 Albert Einstein proposed a new quantum theory of light and explained the photoelectric effect in a landmark paper, for which he received the Nobel Prize in Physics in 1921.

Vadim Lashkaryov discovered *p-n*-junctions in Cu 2 O and silver sulphide protocells in 1941.

Russell Ohl patented the modern junction semiconductor solar cell in 1946 while working on the series of advances that would lead to the transistor.

The first practical photovoltaic cell was publicly demonstrated on 25 April 1954 at Bell Laboratories. The inventors were Daryl Chapin, Calvin Souther Fuller and Gerald Pearson.

Solar cells gained prominence with their incorporation onto the 1958 Vanguard I satellite.

Space Applications

Solar cells were first used in a prominent application when they were proposed and flown on the Vanguard satellite in 1958, as an alternative power source to the primary battery power source. By adding cells to the outside of the body, the mission time could be extended with no major changes to the spacecraft or its power systems. In 1959 the United States launched Explorer 6, featuring large wing-shaped solar arrays, which became a common feature in satellites. These arrays consisted of 9600 Hoffman solar cells.

By the 1960s, solar cells were (and still are) the main power source for most Earth orbiting satellites and a number of probes into the solar system, since they offered the best power-to-weight ratio. However, this success was possible because in the space application, power system costs could be high, because space users had few other power options, and were willing to pay for the best possible cells. The space power market drove the development of higher efficiencies in solar cells up until the National Science Foundation "Research Applied to National Needs" program began to push development of solar cells for terrestrial applications.

In the early 1990s the technology used for space solar cells diverged from the silicon technology used for terrestrial panels, with the spacecraft application shifting to gallium arsenide-based III-V semiconductor materials, which then evolved into the modern III-V multijunction photovoltaic cell used on spacecraft.

Price Reductions

Improvements were gradual over the 1960s. This was also the reason that costs remained high, because space users were willing to pay for the best possible cells, leaving no reason to invest in lower-cost, less-efficient solutions. The price was determined largely by the semiconductor industry; their move to integrated circuits in the 1960s led to the availability of larger boules at lower relative prices. As their price fell, the price of the resulting cells did as well. These effects lowered 1971 cell costs to some $100 per watt.

In late 1969 Elliot Berman joined Exxon's task force which was looking for projects 30 years in the future and in April 1973 he founded Solar Power Corporation, a wholly owned subsidiary of Exxon at that time. The group had concluded that electrical power would be much more expensive by 2000, and felt that this increase in price would make alternative energy sources more attractive. He conducted a market study and concluded that a price per watt of about $20/watt would create significant demand. The team eliminated the steps of polishing the wafers and coating them with an anti-reflective layer, relying on the rough-sawn wafer surface. The team also replaced the expensive materials and hand wiring used in space applications with a printed circuit board on the back, acrylic plastic on the front, and silicone glue between the two, "potting" the cells. Solar cells could be made using cast-off material from the electronics market. By 1973 they announced a product, and SPC convinced Tideland Signal to use its panels to power navigational buoys, initially for the U.S. Coast Guard.

Research and Industrial Production

Research into solar power for terrestrial applications became prominent with the U.S. National Science Foundation's Advanced Solar Energy Research and Development Division within the "Research Applied to National Needs" program, which ran from 1969 to 1977, and funded research on developing solar power for ground electrical power systems. A 1973 conference, the "Cherry Hill Conference", set forth the technology goals required to achieve this goal and outlined an ambitious project for achieving them, kicking off an applied research program that would be ongoing for several decades. The program was eventually taken over by the Energy Research and Development Administration (ERDA), which was later merged into the U.S. Department of Energy.

Following the 1973 oil crisis, oil companies used their higher profits to start (or buy) solar firms, and were for decades the largest producers. Exxon, ARCO, Shell, Amoco (later purchased by BP) and Mobil all had major solar divisions during the 1970s and 1980s. Technology companies also participated, including General Electric, Motorola, IBM, Tyco and RCA.

Declining Costs and Exponential Growth

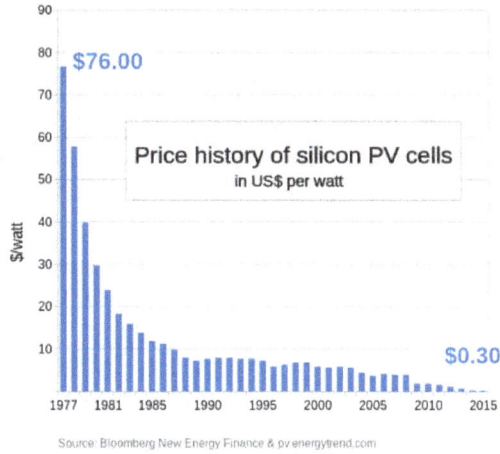

Price per watt history for conventional (c-Si) solar cells since 1977

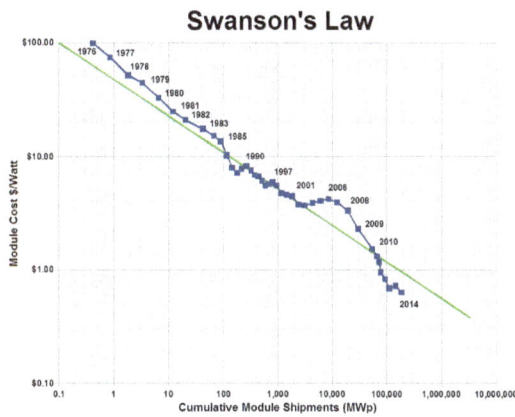

Swanson's law – the learning curve of solar PV

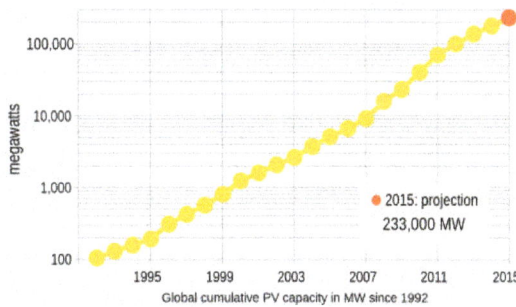

Growth of photovoltaics – Worldwide total installed PV capacity

Adjusting for inflation, it cost $96 per watt for a solar module in the mid-1970s. Process improvements and a very large boost in production have brought that figure down 99 percent, to 68¢ per watt in 2016, according to data from Bloomberg New Energy Finance. Swanson's law is an observation similar to Moore's Law that states that solar cell prices fall 20% for every doubling of industry capacity. It was featured in an article in the British weekly newspaper The Economist.

Further improvements reduced production cost to under $1 per watt, with wholesale costs well under $2. Balance of system costs were then higher than those of the panels. Large commercial arrays could be built, as of 2010, at below $3.40 a watt, fully commissioned.

As the semiconductor industry moved to ever-larger boules, older equipment became inexpensive. Cell sizes grew as equipment became available on the surplus market; ARCO Solar's original panels used cells 2 to 4 inches (50 to 100 mm) in diameter. Panels in the 1990s and early 2000s generally used 125 mm wafers; since 2008 almost all new panels use 156 mm cells. The widespread introduction of flat screen televisions in the late 1990s and early 2000s led to the wide availability of large, high-quality glass sheets to cover the panels.

During the 1990s, polysilicon ("poly") cells became increasingly popular. These cells offer less efficiency than their monosilicon ("mono") counterparts, but they are grown in large vats that reduce cost. By the mid-2000s, poly was dominant in the low-cost panel market, but more recently the mono returned to widespread use.

Manufacturers of wafer-based cells responded to high silicon prices in 2004–2008 with rapid reductions in silicon consumption. In 2008, according to Jef Poortmans, director of IMEC's organic and solar department, current cells use 8–9 grams (0.28–0.32 oz) of silicon per watt of power generation, with wafer thicknesses in the neighborhood of 200 microns. Crystalline silicon panels dominate worldwide markets and are mostly manufactured in China and Taiwan. By late 2011, a drop in European demand due to budgetary turmoil dropped prices for crystalline solar modules to about $1.09 per watt down sharply from 2010. Prices continued to fall in 2012, reaching $0.62/watt by 4Q2012.

Global installed PV capacity reached at least 177 gigawatts in 2014, enough to supply 1 percent of the world's total electricity consumption. Solar PV is growing fastest in Asia, with China and Japan currently accounting for half of worldwide deployment.

Subsidies and Grid Parity

Solar-specific feed-in tariffs vary by country and within countries. Such tariffs encourage the development of solar power projects. Widespread grid parity, the point at which photovoltaic electricity is equal to or cheaper than grid power without subsidies, likely requires advances on all three fronts. Proponents of solar hope to achieve grid parity first in areas with abundant sun and high electricity costs such as in California and Japan. In 2007 BP claimed grid parity for Hawaii and other islands that otherwise use diesel fuel to produce electricity. George W. Bush set 2015 as the date for grid parity in the US. The Photovoltaic Association reported in 2012 that Australia had reached grid parity (ignoring feed in tariffs).

The price of solar panels fell steadily for 40 years, interrupted in 2004 when high subsidies in Germany drastically increased demand there and greatly increased the price of purified silicon (which is used in computer chips as well as solar panels). The recession of 2008 and the onset of Chinese manufacturing caused prices to resume their decline. In the four years after January 2008 prices for solar modules in Germany dropped from €3 to €1 per peak watt. During that same time production capacity surged with an annual growth of more than 50%. China increased market share from 8% in 2008 to over 55% in the last quarter of 2010. In December 2012 the price of Chinese solar panels had dropped to $0.60/Wp (crystalline modules).

Theory

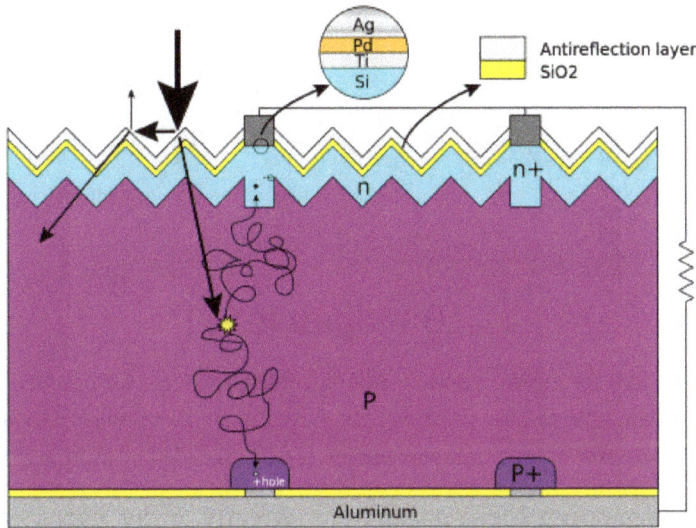

Working mechanism of a solar cell

The solar cell works in several steps:

- Photons in sunlight hit the solar panel and are absorbed by semiconducting materials, such as silicon.

- Electrons are excited from their current molecular/atomic orbital. Once excited an electron can either dissipate the energy as heat and return to its orbital or travel through the cell until it reaches an electrode. Current flows through the material to cancel the potential and this electricity is captured. The chemical bonds of the material are vital for this process to work, and usually silicon is used in two layers, one layer being bonded with boron, the other phosphorus. These layers have different chemical electric charges and subsequently both drive and direct the current of electrons.

- An array of solar cells converts solar energy into a usable amount of direct current (DC) electricity.

- An inverter can convert the power to alternating current (AC).

The most commonly known solar cell is configured as a large-area p–n junction made from silicon.

Efficiency

Solar cell efficiency may be broken down into reflectance efficiency, thermodynamic efficiency, charge carrier separation efficiency and conductive efficiency. The overall efficiency is the product of these individual metrics.

A solar cell has a voltage dependent efficiency curve, temperature coefficients, and allowable shadow angles.

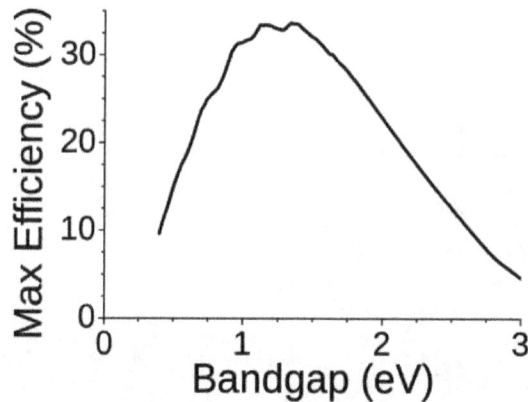

The Shockley-Queisser limit for the theoretical maximum efficiency of a solar cell. Semiconductors with band gap between 1 and 1.5eV, or near-infrared light, have the greatest potential to form an efficient single-junction cell. (The efficiency "limit" shown here can be exceeded by multijunction solar cells.)

Due to the difficulty in measuring these parameters directly, other parameters are substituted: thermodynamic efficiency, quantum efficiency, integrated quantum efficiency, V_{oc} ratio, and fill factor. Reflectance losses are a portion of quantum efficiency under "external quantum efficiency". Recombination losses make up another portion of quantum efficiency, V_{oc} ratio, and fill factor. Resistive losses are predominantly categorized under fill factor, but also make up minor portions of quantum efficiency, V_{oc} ratio.

The fill factor is the ratio of the actual maximum obtainable power to the product of the open circuit voltage and short circuit current. This is a key parameter in evaluating performance. In 2009, typical commercial solar cells had a fill factor > 0.70. Grade B cells were usually between 0.4 and 0.7. Cells with a high fill factor have a low equivalent series resistance and a high equivalent shunt resistance, so less of the current produced by the cell is dissipated in internal losses.

Single p–n junction crystalline silicon devices are now approaching the theoretical limiting power efficiency of 33.7%, noted as the Shockley–Queisser limit in 1961. In the extreme, with an infinite number of layers, the corresponding limit is 86% using concentrated sunlight.

In December 2014, a solar cell achieved a new laboratory record with 46 percent efficiency in a French-German collaboration.

In 2014, three companies broke the record of 25.6% for a silicon solar cell. Panasonic's was the most efficient. The company moved the front contacts to the rear of the panel, eliminating shaded areas. In addition they applied thin silicon films to the (high quality silicon) wafer's front and back to eliminate defects at or near the wafer surface.

In September 2015, the Fraunhofer Institute for Solar Energy Systems (Fraunhofer ISE) announced the achievement of an efficiency above 20% for epitaxial wafer cells. The work on optimizing the atmospheric-pressure chemical vapor deposition (APCVD) in-line production chain was done in collaboration with NexWafe GmbH, a company spun off from Fraunhofer ISE to commercialize production.

For triple-junction thin-film solar cells, the world record is 13.6%, set in June 2015.

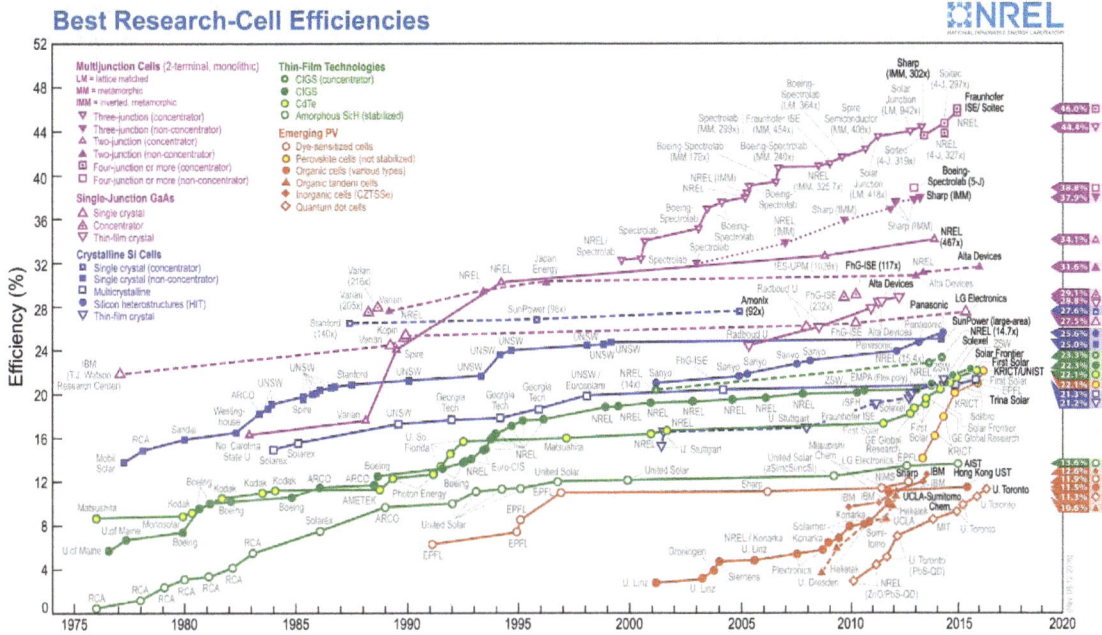

Reported timeline of solar cell energy conversion efficiencies (National Renewable Energy Laboratory)

Materials

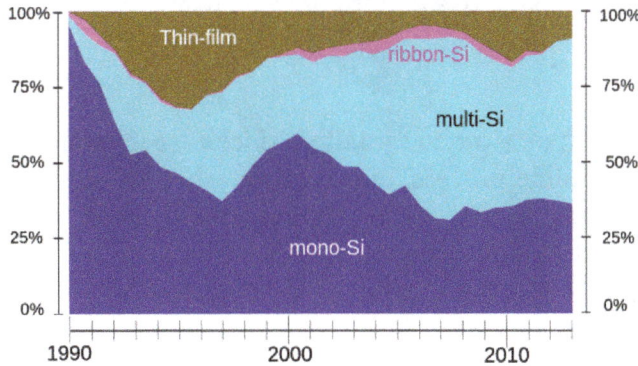

Global market-share in terms of annual production by PV technology since 1990

Solar cells are typically named after the semiconducting material they are made of. These materials must have certain characteristics in order to absorb sunlight. Some cells are designed to handle sunlight that reaches the Earth's surface, while others are optimized for use in space. Solar cells can be made of only one single layer of light-absorbing material (single-junction) or use multiple physical configurations (multi-junctions) to take advantage of various absorption and charge separation mechanisms.

Solar cells can be classified into first, second and third generation cells. The first generation cells—also called conventional, traditional or wafer-based cells—are made of crystalline silicon, the commercially predominant PV technology, that includes materials such as polysilicon and monocrystalline silicon. Second generation cells are thin film solar cells, that include amorphous silicon,

CdTe and CIGS cells and are commercially significant in utility-scale photovoltaic power stations, building integrated photovoltaics or in small stand-alone power system. The third generation of solar cells includes a number of thin-film technologies often described as emerging photovoltaics—most of them have not yet been commercially applied and are still in the research or development phase. Many use organic materials, often organometallic compounds as well as inorganic substances. Despite the fact that their efficiencies had been low and the stability of the absorber material was often too short for commercial applications, there is a lot of research invested into these technologies as they promise to achieve the goal of producing low-cost, high-efficiency solar cells.

Crystalline Silicon

By far, the most prevalent bulk material for solar cells is crystalline silicon (c-Si), also known as "solar grade silicon". Bulk silicon is separated into multiple categories according to crystallinity and crystal size in the resulting ingot, ribbon or wafer. These cells are entirely based around the concept of a p-n junction. Solar cells made of c-Si are made from wafers between 160 and 240 micrometers thick.

Monocrystalline Silicon

Monocrystalline silicon (mono-Si) solar cells are more efficient and more expensive than most other types of cells. The corners of the cells look clipped, like an octagon, because the wafer material is cut from cylindrical ingots, that are typically grown by the Czochralski process. Solar panels using mono-Si cells display a distinctive pattern of small white diamonds.

Epitaxial Silicon

Epitaxial wafers can be grown on a monocrystalline silicon "seed" wafer by atmospheric-pressure CVD in a high-throughput inline process, and then detached as self-supporting wafers of some standard thickness (e.g., 250 μm) that can be manipulated by hand, and directly substituted for wafer cells cut from monocrystalline silicon ingots. Solar cells made with this technique can have efficiencies approaching those of wafer-cut cells, but at appreciably lower cost.

Polycrystalline Silicon

Polycrystalline silicon, or multicrystalline silicon (multi-Si) cells are made from cast square ingots—large blocks of molten silicon carefully cooled and solidified. They consist of small crystals giving the material its typical metal flake effect. Polysilicon cells are the most common type used in photovoltaics and are less expensive, but also less efficient, than those made from monocrystalline silicon.

Ribbon Silicon

Ribbon silicon is a type of polycrystalline silicon—it is formed by drawing flat thin films from molten silicon and results in a polycrystalline structure. These cells are cheaper to make than multi-Si, due to a great reduction in silicon waste, as this approach does not require sawing from ingots. However, they are also less efficient.

Mono-like-multi Silicon (MLM)

This form was developed in the 2000s and introduced commercially around 2009. Also called cast-mono, this design uses polycrystalline casting chambers with small "seeds" of mono material. The result is a bulk mono-like material that is polycrystalline around the outsides. When sliced for processing, the inner sections are high-efficiency mono-like cells (but square instead of "clipped"), while the outer edges are sold as conventional poly. This production method results in mono-like cells at poly-like prices.

Thin Film

Thin-film technologies reduce the amount of active material in a cell. Most designs sandwich active material between two panes of glass. Since silicon solar panels only use one pane of glass, thin film panels are approximately twice as heavy as crystalline silicon panels, although they have a smaller ecological impact (determined from life cycle analysis).

Cadmium Telluride

Cadmium telluride is the only thin film material so far to rival crystalline silicon in cost/watt. However cadmium is highly toxic and tellurium (anion: "telluride") supplies are limited. The cadmium present in the cells would be toxic if released. However, release is impossible during normal operation of the cells and is unlikely during fires in residential roofs. A square meter of CdTe contains approximately the same amount of Cd as a single C cell nickel-cadmium battery, in a more stable and less soluble form.

Copper Indium Gallium Selenide

Copper indium gallium selenide (CIGS) is a direct band gap material. It has the highest efficiency (~20%) among all commercially significant thin film materials. Traditional methods of fabrication involve vacuum processes including co-evaporation and sputtering. Recent developments at IBM and Nanosolar attempt to lower the cost by using non-vacuum solution processes.

Silicon Thin Film

Silicon thin-film cells are mainly deposited by chemical vapor deposition (typically plasma-enhanced, PE-CVD) from silane gas and hydrogen gas. Depending on the deposition parameters, this can yield amorphous silicon (a-Si or a-Si:H), protocrystalline silicon or nanocrystalline silicon (nc-Si or nc-Si:H), also called microcrystalline silicon.

Amorphous silicon is the most well-developed thin film technology to-date. An amorphous silicon (a-Si) solar cell is made of non-crystalline or microcrystalline silicon. Amorphous silicon has a higher bandgap (1.7 eV) than crystalline silicon (c-Si) (1.1 eV), which means it absorbs the visible part of the solar spectrum more strongly than the higher power density infrared portion of the spectrum. The production of a-Si thin film solar cells uses glass as a substrate and deposits a very thin layer of silicon by plasma-enhanced chemical vapor deposition (PECVD).

Protocrystalline silicon with a low volume fraction of nanocrystalline silicon is optimal for high open circuit voltage. Nc-Si has about the same bandgap as c-Si and nc-Si and a-Si can advantageously be combined in thin layers, creating a layered cell called a tandem cell. The top cell in a-Si absorbs the visible light and leaves the infrared part of the spectrum for the bottom cell in nc-Si.

Gallium Arsenide Thin Film

The semiconductor material Gallium arsenide (GaAs) is also used for single-crystalline thin film solar cells. Although GaAs cells are very expensive, they hold the world's record in efficiency for a single-junction solar cell at 28.8%. GaAs is more commonly used in multijunction photovoltaic cells for concentrated photovoltaics (CPV, HCPV) and for solar panels on spacecrafts, as the industry favours efficiency over cost for space-based solar power.

Multijunction Cells

Dawn's 10 kW triple-junction gallium arsenide solar array at full extension

Multi-junction cells consist of multiple thin films, each essentially a solar cell grown on top of another, typically using metalorganic vapour phase epitaxy. Each layers has a different band gap energy to allow it to absorb electromagnetic radiation over a different portion of the spectrum. Multi-junction cells were originally developed for special applications such as satellites and space exploration, but are now used increasingly in terrestrial concentrator photovoltaics (CPV), an emerging technology that uses lenses and curved mirrors to concentrate sunlight onto small but highly efficient multi-junction solar cells. By concentrating sunlight up to a thousand times, *High concentrated photovoltaics (HCPV)* has the potential to outcompete conventional solar PV in the future.

Tandem solar cells based on monolithic, series connected, gallium indium phosphide (GaInP), gallium arsenide (GaAs), and germanium (Ge) p–n junctions, are increasing sales, despite cost pressures. Between December 2006 and December 2007, the cost of 4N gallium metal rose from about $350 per kg to $680 per kg. Additionally, germanium metal prices have risen substantially to $1000–1200 per kg this year. Those materials include gallium (4N, 6N and 7N Ga), arsenic (4N, 6N and 7N) and germanium, pyrolitic boron nitride (pBN) crucibles for growing crystals, and boron oxide, these products are critical to the entire substrate manufacturing industry.

A triple-junction cell, for example, may consist of the semiconductors: GaAs, Ge, and GaInP 2. Triple-junction GaAs solar cells were used as the power source of the Dutch four-time World Solar Challenge winners Nuna in 2003, 2005 and 2007 and by the Dutch solar cars Solutra (2005),

Twente One (2007) and 21Revolution (2009). GaAs based multi-junction devices are the most efficient solar cells to date. On 15 October 2012, triple junction metamorphic cells reached a record high of 44%.

Research in Solar Cells

Perovskite Solar Cells

Perovskite solar cells are solar cells that include a perovskite-structured material as the active layer. Most commonly, this is a solution-processed hybrid organic-inorganic tin or lead halide based material. Efficiencies have increased from below 5% at their first usage in 2009 to over 20% in 2014, making them a very rapidly advancing technology and a hot topic in the solar cell field. Perovskite solar cells are also forecast to be extremely cheap to scale up, making them a very attractive option for commercialisation.

Liquid Inks

In 2014, researchers at California NanoSystems Institute discovered using kesterite and perovskite improved electric power conversion efficiency for solar cells.

Upconversion and Downconversion

Photon upconversion is the process of using two low-energy (*e.g.*, infrared) photons to produce one higher energy photon; downconversion is the process of using one high energy photon (*e.g.,*, ultraviolet) to produce two lower energy photons. Either of these techniques could be used to produce higher efficiency solar cells by allowing solar photons to be more efficiently used. The difficulty, however, is that the conversion efficiency of existing phosphors exhibiting up- or downconversion is low, and is typically narrow band.

One upconversion technique is to incorporate lanthanide-doped materials ($Er3+$, $Yb3+$, $Ho3+$ or a combination), taking advantage of their luminescence to convert infrared radiation to visible light. Upconversion process occurs when two infrared photons are absorbed by rare-earth ions to generate a (high-energy) absorbable photon. As example, the energy transfer upconversion process (ETU), consists in successive transfer processes between excited ions in the near infrared. The upconverter material could be placed below the solar cell to absorb the infrared light that passes through the silicon. Useful ions are most commonly found in the trivalent state. $Er+$ ions have been the most used. $Er3+$ ions absorb solar radiation around 1.54 μm. Two $Er3+$ ions that have absorbed this radiation can interact with each other through an upconversion process. The excited ion emits light above the Si bandgap that is absorbed by the solar cell and creates an additional electron–hole pair that can generate current. However, the increased efficiency was small. In addition, fluoroindate glasses have low phonon energy and have been proposed as suitable matrix doped with $Ho3+$ ions.

Light-absorbing Dyes

Dye-sensitized solar cells (DSSCs) are made of low-cost materials and do not need elaborate manufacturing equipment, so they can be made in a DIY fashion. In bulk it should be significantly less expensive than older solid-state cell designs. DSSC's can be engineered into flexible sheets and although its conversion efficiency is less than the best thin film cells, its price/performance ratio may be high enough to allow them to compete with fossil fuel electrical generation.

Typically a ruthenium metalorganic dye (Ru-centered) is used as a monolayer of light-absorbing material. The dye-sensitized solar cell depends on a mesoporous layer of nanoparticulate titanium dioxide to greatly amplify the surface area (200–300 m^2/g TiO 2, as compared to approximately 10 m^2/g of flat single crystal). The photogenerated electrons from the light absorbing dye are passed on to the n-type TiO 2 and the holes are absorbed by an electrolyte on the other side of the dye. The circuit is completed by a redox couple in the electrolyte, which can be liquid or solid. This type of cell allows more flexible use of materials and is typically manufactured by screen printing or ultrasonic nozzles, with the potential for lower processing costs than those used for bulk solar cells. However, the dyes in these cells also suffer from degradation under heat and UV light and the cell casing is difficult to seal due to the solvents used in assembly. The first commercial shipment of DSSC solar modules occurred in July 2009 from G24i Innovations.

Quantum Dots

Quantum dot solar cells (QDSCs) are based on the Gratzel cell, or dye-sensitized solar cell architecture, but employ low band gap semiconductor nanoparticles, fabricated with crystallite sizes small enough to form quantum dots (such as CdS, CdSe, Sb 2S 3, PbS, etc.), instead of organic or organometallic dyes as light absorbers. QD's size quantization allows for the band gap to be tuned by simply changing particle size. They also have high extinction coefficients and have shown the possibility of multiple exciton generation.

In a QDSC, a mesoporous layer of titanium dioxide nanoparticles forms the backbone of the cell, much like in a DSSC. This TiO 2 layer can then be made photoactive by coating with semiconductor quantum dots using chemical bath deposition, electrophoretic deposition or successive ionic layer adsorption and reaction. The electrical circuit is then completed through the use of a liquid or solid redox couple. The efficiency of QDSCs has increased to over 5% shown for both liquid-junction and solid state cells. In an effort to decrease production costs, the Prashant Kamat research group demonstrated a solar paint made with TiO 2 and CdSe that can be applied using a one-step method to any conductive surface with efficiencies over 1%.

Organic/Polymer Solar Cells

Organic solar cells and polymer solar cells are built from thin films (typically 100 nm) of organic semiconductors including polymers, such as polyphenylene vinylene and small-molecule compounds like copper phthalocyanine (a blue or green organic pigment) and carbon fullerenes and fullerene derivatives such as PCBM.

They can be processed from liquid solution, offering the possibility of a simple roll-to-roll printing process, potentially leading to inexpensive, large-scale production. In addition, these cells could be beneficial for some applications where mechanical flexibility and disposability are important. Current cell efficiencies are, however, very low, and practical devices are essentially non-existent.

Energy conversion efficiencies achieved to date using conductive polymers are very low compared to inorganic materials. However, Konarka Power Plastic reached efficiency of 8.3% and organic tandem cells in 2012 reached 11.1%.

The active region of an organic device consists of two materials, one electron donor and one electron acceptor. When a photon is converted into an electron hole pair, typically in the donor material, the charges tend to remain bound in the form of an exciton, separating when the exciton diffuses to the donor-acceptor interface, unlike most other solar cell types. The short exciton diffusion lengths of most polymer systems tend to limit the efficiency of such devices. Nanostructured interfaces, sometimes in the form of bulk heterojunctions, can improve performance.

In 2011, MIT and Michigan State researchers developed solar cells with a power efficiency close to 2% with a transparency to the human eye greater than 65%, achieved by selectively absorbing the ultraviolet and near-infrared parts of the spectrum with small-molecule compounds. Researchers at UCLA more recently developed an analogous polymer solar cell, following the same approach, that is 70% transparent and has a 4% power conversion efficiency. These lightweight, flexible cells can be produced in bulk at a low cost and could be used to create power generating windows.

In 2013, researchers announced polymer cells with some 3% efficiency. They used block copolymers, self-assembling organic materials that arrange themselves into distinct layers. The research focused on P3HT-b-PFTBT that separates into bands some 16 nanometers wide.

Adaptive Cells

Adaptive cells change their absorption/reflection characteristics depending to respond to environmental conditions. An adaptive material responds to the intensity and angle of incident light. At the part of the cell where the light is most intense, the cell surface changes from reflective to adaptive, allowing the light to penetrate the cell. The other parts of the cell remain reflective increasing the retention of the absorbed light within the cell.

In 2014 a system that combined an adaptive surface with a glass substrate that redirect the absorbed to a light absorber on the edges of the sheet. The system also included an array of fixed lenses/mirrors to concentrate light onto the adaptive surface. As the day continues, the concentrated light moves along the surface of the cell. That surface switches from reflective to adaptive when the light is most concentrated and back to reflective after the light moves along.

Manufacture

Solar cells share some of the same processing and manufacturing techniques as other semiconductor devices. However, the stringent requirements for cleanliness and quality control of semiconductor fabrication are more relaxed for solar cells, lowering costs.

Polycrystalline silicon wafers are made by wire-sawing block-cast silicon ingots into 180 to 350 micrometer wafers. The wafers are usually lightly p-type-doped. A surface diffusion of n-type dopants is performed on the front side of the wafer. This forms a p–n junction a few hundred nanometers below the surface.

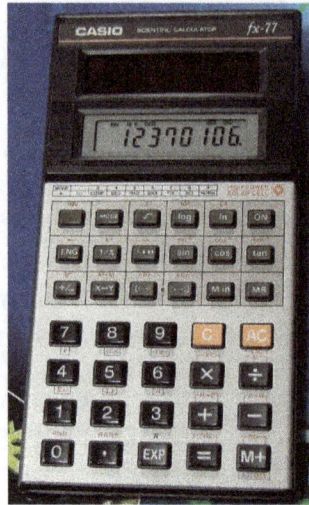

Early solar-powered calculator

Anti-reflection coatings are then typically applied to increase the amount of light coupled into the solar cell. Silicon nitride has gradually replaced titanium dioxide as the preferred material, because of its excellent surface passivation qualities. It prevents carrier recombination at the cell surface. A layer several hundred nanometers thick is applied using PECVD. Some solar cells have textured front surfaces that, like anti-reflection coatings, increase the amount of light reaching the wafer. Such surfaces were first applied to single-crystal silicon, followed by multicrystalline silicon somewhat later.

A full area metal contact is made on the back surface, and a grid-like metal contact made up of fine "fingers" and larger "bus bars" are screen-printed onto the front surface using a silver paste. This is an evolution of the so-called "wet" process for applying electrodes, first described in a US patent filed in 1981 by Bayer AG. The rear contact is formed by screen-printing a metal paste, typically aluminium. Usually this contact covers the entire rear, though some designs employ a grid pattern. The paste is then fired at several hundred degrees Celsius to form metal electrodes in ohmic contact with the silicon. Some companies use an additional electro-plating step to increase efficiency. After the metal contacts are made, the solar cells are interconnected by flat wires or metal ribbons, and assembled into modules or "solar panels". Solar panels have a sheet of tempered glass on the front, and a polymer encapsulation on the back.

Manufacturers and Certification

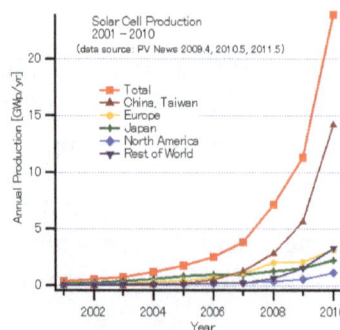

Solar cell production by region

National Renewable Energy Laboratory tests and validates solar technologies. Three reliable groups certify solar equipment: UL and IEEE (both U.S. standards) and IEC.

Solar cells are manufactured in volume in Japan, Germany, China, Taiwan, Malaysia and the United States, whereas Europe, China, the U.S., and Japan have dominated (94% or more as of 2013) in installed systems. Other nations are acquiring significant solar cell production capacity.

Global PV cell/module production increased by 10% in 2012 despite a 9% decline in solar energy investments according to the annual "PV Status Report" released by the European Commission's Joint Research Centre. Between 2009 and 2013 cell production has quadrupled.

China

Due to heavy government investment, China has become the dominant force in solar cell manufacturing. Chinese companies produced solar cells/modules with a capacity of ~23 GW in 2013 (60% of global production).

Malaysia

In 2014, Malaysia was the world's third largest manufacturer of photovoltaics equipment, behind China and the European Union.

United States

Solar cell production in the U.S. has suffered due to the global financial crisis, but recovered partly due to the falling price of quality silicon.

Biomass Briquettes

Briquette made by a Ruf briquetter out of hay

Straw or hay briquettes

Ogatan, Japanese charcoal briquettes made from sawdust briquettes *(Ogalite)*.

Quick Grill Briquette made from coconut shell

Biomass briquettes are a biofuel substitute to coal and charcoal. Briquettes are mostly used in the developing world, where cooking fuels are not as easily available. There has been a move to the use of briquettes in the developed world, where they are used to heat industrial boilers in order to produce electricity from steam. The briquettes are cofired with coal in order to create the heat supplied to the boiler.

Composition and Production

Biomass briquettes, mostly made of green waste and other organic materials, are commonly used for electricity generation, heat, and cooking fuel. These compressed compounds contain various organic materials, including rice husk, bagasse, ground nut shells, municipal solid waste, agricultural waste. The composition of the briquettes varies by area due to the availability of raw materials. The raw materials are gathered and compressed into briquette in order to burn longer and make transportation of the goods easier. These briquettes are very different from charcoal because they do not have large concentrations of carbonaceous substances and added materials. Compared to fossil fuels, the briquettes produce low net total greenhouse gas emissions because the materials used are already a part of the carbon cycle.

One of the most common variables of the biomass briquette production process is the way the biomass is dried out. Manufacturers can use torrefaction, carbonization, or varying degrees of pyrolysis. Researchers concluded that torrefaction and carbonization are the most efficient forms of drying out biomass, but the use of the briquette determines which method should be used.

Compaction is another factor affecting production. Some materials burn more efficiently if compacted at low pressures, such as corn stover grind. Other materials such as wheat and barleystraw require high amounts of pressure to produce heat. There are also different press technologies that can be used. A piston press is used to create solid briquettes for a wide array of purposes. Screw extrusion is used to compact biomass into loose, homogeneous briquettes that are substituted for coal in cofiring. This technology creates a toroidal, or doughnut-like, briquette. The hole in the center of the briquette allows for a larger surface area, creating a higher combustion rate.

History

People have been using biomass briquettes in Nepal since before recorded history. Though inefficient, the burning of loose biomass created enough heat for cooking purposes and keeping warm. The first commercial production plant was created in 1982 and produced almost 900 metric tons of biomass. In 1984, factories were constructed that incorporated vast improvements on efficiency and the quality of briquettes. They used a combination of rice husks and molasses. The King Mahendra Trust for Nature Conservation (KMTNC) along with the Institute for Himalayan Conservation (IHC) created a mixture of coal and biomass in 2000 using a unique rolling machine.

Japanese Ogalite

In 1925, Japan independently started developing technology to harness the energy from sawdust briquettes, known as *"Ogalite"*. Between 1964 and 1969, Japan increased production fourfold by incorporating screw press and piston press technology. The member enterprise of 830 or more existed in the 1960s. The new compaction techniques incorporated in these machines made briquettes of higher quality than those in Europe. As a result, European countries bought the licensing agreements and now manufacture Japanese designed machines.

Cofiring

Cofiring relates to the combustion of two different types of materials. The process is primarily used to decrease CO_2 emissions despite the resulting lower energy efficiency and higher variable cost. The combination of materials usually contains a high carbon emitting substance such as coal and a lesser CO_2 emitting material such as biomass. Even though CO_2 will still be emitted through the combustion of biomass, the net carbon emitted is nearly negligible. This is due to the fact that the material gathered for the composition of the briquettes are still contained in the carbon cycle whereas fossil fuel combustion releases CO_2 that has been sequestered for millennia. Boilers in power plants are traditionally heated by the combustion of coal, but if cofiring were to be implemented, then the CO_2 emissions would decrease while still maintaining the heat inputted to the boiler. Implementing cofiring would require few modifications to the current characteristics to power plants, as only the fuel for the boiler would be altered. A moderate investment would be required for implementing biomass briquettes into the combustion process.

Cofiring is considered the most cost-efficient means of biomass. A higher combustion rate will occur when cofiring is implemented in a boiler when compared to burning only biomass. The compressed biomass is also much easier to transport since it is more dense, therefore allowing more biomass to be transported per shipment when compared to loose biomass. Some sources agree that a near-term solution for the greenhouse gas emission problem may lie in cofiring.

Compared to Coal

The use of biomass briquettes has been steadily increasing as industries realize the benefits of decreasing pollution through the use of biomass briquettes. Briquettes provide higher calorific value per dollar than coal when used for firing industrial boilers. Along with higher calorific value, biomass briquettes on average saved 30–40% of boiler fuel cost. But other sources suggest that cofiring is more expensive due to the widespread availability of coal and its low cost. However, in the long run, briquettes can only limit the use of coal to a small extent, but it is increasingly being pursued by industries and factories all over the world. Both raw materials can be produced or mined domestically in the United States, creating a fuel source that is free from foreign dependence and less polluting than raw fossil fuel incineration.

Environmentally, the use of biomass briquettes produces much fewer greenhouse gases, specifically, 13.8% to 41.7% CO_2 and NO_x. There was also a reduction from 11.1% to 38.5% in SO 2 emissions when compared to coal from three different leading producers, EKCC Coal, Decanter Coal, and Alden Coal. Biomass briquettes are also fairly resistant to water degradation, an improvement over the difficulties encountered with the burning of wet coal. However, the briquettes are best used only as a supplement to coal. The use of cofiring creates an energy that is not as high as pure coal, but emits fewer pollutants and cuts down on the release of previously sequestered carbon. The continuous release of carbon and other greenhouse gasses into the atmosphere leads to an increase in global temperatures. The use of cofiring does not stop this process but decreases the relative emissions of coal power plants.

Use in Developing World

The Legacy Foundation has developed a set of techniques to produce biomass briquettes through artisanal production in rural villages that can be used for heating and cooking. These techniques were recently pioneered by Virunga National Park in eastern Democratic Republic of Congo, following the massive destruction of the mountain gorilla habitat for charcoal.

Pangani, Tanzania, is an area covered in coconut groves. After harvesting the meat of the coconut, the indigenous people would litter the ground with the husks, believing them to be useless. The husks later became a profit center after it was discovered that coconut husks are well suited to be the main ingredient in bio briquettes. This alternative fuel mixture burns incredibly efficiently and leaves little residue, making it a reliable source for cooking in the undeveloped country. The developing world has always relied on the burning biomass due it its low cost and availability anywhere there is organic material. The briquette production only improves upon the ancient practice by increasing the efficiency of pyrolysis.

Two major components of the developing world are China and India. The economies are rapidly increasing due to cheap ways of harnessing electricity and emitting large amounts of carbon dioxide. The Kyoto Protocol attempted to regulate the emissions of the three different worlds, but there were disagreements as to which country should be penalized for emissions based on its previous and future emissions. The United States has been the largest emitter but China has recently become the largest per capita. The United States had emitted a rigorous amount of carbon dioxide during its development and the developing nations argue that they should not be forced to meet the requirements. At the lower end, the undeveloped nations believe that they have little responsibility for what has been done to the carbon dioxide levels. The major use of biomass briquettes in India, is in industrial applications usually to produce steam. A lot of conversions of boilers from FO to biomass briquettes have happened over the past decade. A vast majority of those projects are registered under CDM (Kyoto Protocol), which allows for users to get carbon credits.

The use of biomass briquettes is strongly encouraged by issuing carbon credits. One carbon credit is equal to one free ton of carbon dioxide to be emitted into the atmosphere. India has started to replace charcoal with biomass briquettes in regards to boiler fuel, especially in the southern parts of the country because the biomass briquettes can be created domestically, depending on the availability of land. Therefore, constantly rising fuel prices will be less influential in an economy if sources of fuel can be easily produced domestically. Lehra Fuel Tech Pvt Ltd is approved by Indian Renewable Energy Development Agency (IREDA), is one of the largest briquetting machine manufacturers from Ludhiana, India.

In the African Great Lakes region, work on biomass briquette production has been spearheaded by a number of NGOs with GVEP (Global Village Energy Partnership) taking a lead in promoting briquette products and briquette entrepreneurs in the three Great Lakes countries; namely, Kenya, Uganda and Tanzania. This has been achieved by a five-year EU and Dutch government sponsored project called DEEP EA (Developing Energy Enterprises Project East Africa) . The main feed stock for briquettes in the East African region has mainly been charcoal dust although alternative like sawdust, bagasse, coffee husks and rice husks have also been used.

Use in Developed World

Coal is the largest carbon dioxide emitter per unit area when it comes to electricity generation. It is also the most common ingredient in charcoal. There has been a recent push to replace the burning of fossil fuels with biomass. The replacement of this nonrenewable resource with biological waste would lower the carbon footprint of grill owners and lower the overall pollution of the world. Citizens are also starting to manufacture briquettes at home. The first machines would create briquettes for homeowners out of compressed sawdust, however, current machines allow for briquette production out of any sort of dried biomass.

Arizona has also taken initiative to turn waste biomass into a source of energy. Waste cotton and pecan material used to provide a nesting ground for bugs that would destroy the new crops in the spring. To stop this problem farmers buried the biomass, which quickly led to soil degradation. These materials were discovered to be a very efficient source of energy and took care of issues that had plagued farms.

The United States Department of Energy has financed several projects to test the viability of biomass briquettes on a national scale. The scope of the projects is to increase the efficiency of gasifiers as well as produce plans for production facilities.

Criticism

Biomass is composed of organic materials, therefore, large amounts of land are required to produce the fuel. Critics argue that the use of this land should be utilized for food distribution rather than crop degradation. Also, climate changes may cause a harsh season, where the material extracted will need to be swapped for food rather than energy. The assumption is that the production of biomass decreases the food supply, causing an increase in world hunger by extracting the organic materials such as corn and soybeans for fuel rather than food.

The cost of implementing a new technology such as biomass into the current infrastructure is also high. The fixed costs with the production of biomass briquettes are high due to the new undeveloped technologies that revolve around the extraction, production and storage of the biomass. Technologies regarding extraction of oil and coal have been developing for decades, becoming more efficient with each year. A new undeveloped technology regarding fuel utilization that has no infrastructure built around makes it nearly impossible to compete in the current market.

References

- Ahmad Y Hassan, Donald Routledge Hill (1986). Islamic Technology: An illustrated history, p. 54. Cambridge University Press. ISBN 0-521-42239-6.

- Morthorst, Poul Erik; Redlinger, Robert Y.; Andersen, Per (2002). Wind energy in the 21st century: economics, policy, technology and the changing electricity industry. Houndmills, Basingstoke, Hampshire: Palgrave/ UNEP. ISBN 0-333-79248-3.

- Gevorkian, Peter (2007). Sustainable energy systems engineering: the complete green building design resource. McGraw Hill Professional. ISBN 978-0-07-147359-0.

- Tsokos, K. A. (28 January 2010). Physics for the IB Diploma Full Colour. Cambridge University Press. ISBN 978-0-521-13821-5.

- Williams, Neville (2005). Chasing the Sun: Solar Adventures Around the World. New Society Publishers. p. 84. ISBN 9781550923124.

- Kim, D.S.; et al. (18 May 2003). "String ribbon silicon solar cells with 17.8% efficiency" (PDF). Proceedings of 3rd World Conference on Photovoltaic Energy Conversion, 2003. 2: 1293–1296. ISBN 4-9901816-0-3.

- Pearce, J.; Lau, A. (2002). "Net Energy Analysis for Sustainable Energy Production from Silicon Based Solar Cells". Solar Energy (PDF). p. 181. doi:10.1115/SED2002-1051. ISBN 0-7918-1689-3.

- Yablonovitch, Eli; Miller, Owen D.; Kurtz, S. R. (2012). "The opto-electronic physics that broke the efficiency limit in solar cells". 2012 38th IEEE Photovoltaic Specialists Conference. p. 001556. doi:10.1109/PVSC.2012.6317891. ISBN 978-1-4673-0066-7.

- Morten Lund (30 May 2016). "Dansk firma sætter prisbelønnet selvhejsende kran i serieproduktion". Ingeniøren. Retrieved 30 May 2016.

- Tom Gray (11 March 2013). "Fact check: About those 'abandoned' turbines ...". American Wind Energy Association. Retrieved 30 May 2016.

- Janz, Stefan; Reber, Stefan (14 September 2015). "20% Efficient Solar Cell on EpiWafer". Fraunhofer ISE. Retrieved October 15, 2015.

- Rachow, Thomas; Heinz, Friedemann; Steinhauser, Bernd; Janz, Stefan; Reber, Stefan (2014). "Epitaxial n- and p-type Emitters for High Efficiency Solar Cell Concepts". Journal of Energy and Power Engineering. 8. David Publishing. pp. 1371–1377. Retrieved October 15, 2015.

- "Special Operations Summit Little Creek – Wind Power: "Small and Micro Wind"...A Position of Power I". Tactical Defense Media's DoD Power, Energy & Propulsion Q2 2012, tacticaldefensemedia.com. Retrieved 2015-10-20.

- Wittrup, Sanne. "Power from Vestas' giant turbine" (in Danish. English translation). Ingeniøren, 28 January 2014. Retrieved 28 January 2014.

- "Technology Roadmap: Solar Photovoltaic Energy" (PDF). IEA. 2014. Archived from the original on 7 October 2014. Retrieved 7 October 2014.

- "Photovoltaic System Pricing Trends – Historical, Recent, and Near-Term Projections, 2014 Edition" (PDF). NREL. 22 September 2014. p. 4. Archived from the original on 29 March 2015.

- "French-German collaborators claim solar cell efficiency world record". EE Times Europe. 2 December 2014. Retrieved 3 December 2014.

- PV Status Report 2013 | Renewable Energy Mapping and Monitoring in Europe and Africa (REMEA). Iet.jrc.ec.europa.eu (11 April 2014). Retrieved on 20 April 2014.

Permissions

All chapters in this book are published with permission under the Creative Commons Attribution Share Alike License or equivalent. Every chapter published in this book has been scrutinized by our experts. Their significance has been extensively debated. The topics covered herein carry significant information for a comprehensive understanding. They may even be implemented as practical applications or may be referred to as a beginning point for further studies.

We would like to thank the editorial team for lending their expertise to make the book truly unique. They have played a crucial role in the development of this book. Without their invaluable contributions this book wouldn't have been possible. They have made vital efforts to compile up to date information on the varied aspects of this subject to make this book a valuable addition to the collection of many professionals and students.

This book was conceptualized with the vision of imparting up-to-date and integrated information in this field. To ensure the same, a matchless editorial board was set up. Every individual on the board went through rigorous rounds of assessment to prove their worth. After which they invested a large part of their time researching and compiling the most relevant data for our readers.

The editorial board has been involved in producing this book since its inception. They have spent rigorous hours researching and exploring the diverse topics which have resulted in the successful publishing of this book. They have passed on their knowledge of decades through this book. To expedite this challenging task, the publisher supported the team at every step. A small team of assistant editors was also appointed to further simplify the editing procedure and attain best results for the readers.

Apart from the editorial board, the designing team has also invested a significant amount of their time in understanding the subject and creating the most relevant covers. They scrutinized every image to scout for the most suitable representation of the subject and create an appropriate cover for the book.

The publishing team has been an ardent support to the editorial, designing and production team. Their endless efforts to recruit the best for this project, has resulted in the accomplishment of this book. They are a veteran in the field of academics and their pool of knowledge is as vast as their experience in printing. Their expertise and guidance has proved useful at every step. Their uncompromising quality standards have made this book an exceptional effort. Their encouragement from time to time has been an inspiration for everyone.

The publisher and the editorial board hope that this book will prove to be a valuable piece of knowledge for students, practitioners and scholars across the globe.

Index